Badgers

James Lowen

BLOOMSBURY WILDLIFE
Bloomsbury Publishing Plc
50 Bedford Square, London, WC1B 3DP, UK

BLOOMSBURY, BLOOMSBURY WILDLIFE and the Diana logo are trademarks of Bloomsbury Publishing Plc

First published in in the United Kingdom 2016

British Library Cataloguing-in-Publication Data

A catalogue record for this book is available from the British Library

Library of Congress Cataloguing-in-Publication data has been applied for

ISBN: PB: 978-1-4729-7174-6; ePDF: 978-1-4729-2781-1; ePub: 978-1-4729-2182-6

2 4 6 8 10 9 7 5 3 1

Design by Susan McIntyre
Printed and bound in India by Replika Press Pvt. Ltd.

Contents

Meet the Badgers

Beloved by generations of children entranced by Kenneth Grahame's book *The Wind in the Willows*, few mammals appear as mysterious as the Badger. Because of its nocturnal lifestyle, not many of us have ever been lucky enough to even glimpse its monochrome form as it sniffs and bustles its way through woodland or across pasture. The closest we believe we get to these determined mammals is through their cultural proxies – yet most of us live far nearer to a Badger group than we might think.

Whether we have seen one or not, we all know what a Badger – formally, the Eurasian Badger – looks like. A head striped with black and white has made the Badger iconic. It is the face of marketing campaigns and logos, from the UK's Wildlife Trusts, to beer and ink. As an avatar, brand, book illustration or newspaper photograph, the Badger is familiar to us all. Yet few of us have ever seen one alive.

In general the only time we clap eyes on a Badger is when we see a cylindrical, motionless lump on the roadside – a victim of our motorised existence. We lead separate lives, our paths crossing only in death. The Badger's dark world seems largely parallel to ours.

Above: The Badger's iconic face features in advertising campaigns and on logos.

Opposite: The flaring stripes of the Badger's head pattern render the mammal unmistakable.

Below: A typical view of a Badger at night, snuffling through woodland.

This seems bizarre when we have cohabited in Europe for tens of thousands of years – and the Badger has made its mark in our landscape and language, our place names and culture.

There is much about the Badger that we admire and with which we even feel a certain communion. Stoic when faced with adversity and tenacious in defending family and home, the Badger is a battler. Somehow Badgers survive everything humans throw at them, whether this is altering the environment, baiting them with dogs or persecuting them for their role in transmitting disease. We respect Badgers and identify with them – yet we also hound them. We know *of* the Badger, yet cannot claim to know *it*. We are ignorant of the Badger's way of life, and confused by how we are supposed to feel about it. This book seeks to change all that.

A scientific family tree

Above: Carl Linnaeus, the Swedish scientist who created the current system of naming and classifying plants, animals and fungi.

Even experts can get things wrong. The grandfather of taxonomy – the science of classifying creatures – was a Swedish naturalist called Carl Linnaeus. In the mid-18th century, Linnaeus completed the first ever attempt to classify and name all living life forms. Many of his allocations have stood the test of time and are still used today. One notable exception, however, was what we know today as the Eurasian Badger *Meles meles*, which Linnaeus considered to be a form of bear!

With its shaggy coat, long claws and bumbling gait, there is admittedly something bear-like about the Badger. But the bumbling Brock is actually part of the mustelid or weasel family, whose European members include the Otter, Stoat and Pine Marten.

Mustelidae are the largest family in the order Carnivora, which comprises carnivorous mammals. Mustelids are thought to have first evolved 40 million years ago, and today's representatives are descended from mammals that appeared 15 million years ago. To put those dates in perspective, the precursors of our own species, *Homo sapiens*, emerged from an ape-like ancestor as recently as 5–7 million years ago.

Above: In a rare (but understandable) mistake, Carl Linnaeus classified the Badger (right) as a relative of the Brown Bear (left).

Species

Panthera pardus (Leopard)
Mephitis mephitis (Skunk)
Meles meles (Eurasian badger)
Canis familiaris (Domestic dog)
Canis lupus (Wolf)

Genus

Panthera
Mephitis
Meles
Canis

Family

Felidae
Mustelidae
Canidae

Order

Carnivora

Left: Badgers are a branch within the mustelid (or weasel) family, which has been around in its current form for 15 million years – more than twice as long as our own species, *Homo sapiens*.

What's in a name?

There are different suggestions for the origin of the word 'badger'. It could come from the French *bêcheur*, meaning 'digger'. However, the French word for the Badger, *blaireau*, means 'corn hoarder', presumably in recognition of one of the mammal's favourite foods. Alternatively, the name may be derived from 'badge' in recognition of the mammal's banded face – its distinctive emblem amid the ink of night.

Across Britain there are also 'country' names for the Badger. The best known is Brock, which is often used in children's stories and comes from *broc*, a word of Celtic origin and a reference to the mammal's largely grey colouration. There are perhaps 140 Anglo-Saxon place names with their origin in *broc*, from Brockley to Brockenhurst. Another common moniker, particularly in northern England, is Pate. Rarer are Grey and Bawson. The first is presumably a reference to the coat's apparent colour, while the second may have evolved from an Old French word, *bauçant*, meaning 'pied'.

In German, Badger is *Dachs*; in Dutch, it is *das*. The Latin root for Badger, *taxo*, lives on in the mammal's names in Portugal (*teixugo*), Italian (*tasso*), Spanish (*tejón*) and Catalan (*toixó*). The Norwegian *grevling* honours the Badger's proficiency at digging, with a similar etymology in Danish and Swedish. Further east Badgers are ascribed names of contrasting resonance: *vjedhulle* in Albanian (meaning 'thief') and *huan* in Chinese (a homophone for 'happiness').

Above: Possible origins of the word 'Badger' include the badge of its banded face (above) or the French *bêcheur*, meaning 'digger' (right).

Below: A colloquial English name for Badger, Bawson, may derive from an Old French word for 'pied', *bauçant*.

The UK's mustelids

Mustelids roam across every continent bar Antarctica and Australasia. Most are agile and active mammals, always alert to the next hunting opportunity. They have evolved to occupy varied ecological niches, from wetlands to woodland and tunnels to trees. As befits members of the order Carnivora, most are skilful and strong predators. Stoats, for example, are not deterred by their favourite prey – Rabbits – being twice their size. Sea Otters are one of the few mammals to have learnt how to use tools to access their food, in their case stones to break open shellfish.

With the exception of the Red Fox and Scottish Wildcat, all of the UK's carnivores are mustelids. They occur in a variety of shapes and sizes.

- **Badger** The UK's largest mustelid is widespread and lives underground in setts. Although rarely seen, it is as familiar to us as a Fox, largely because of its place in British culture.

- **Otter** With webbed feet, muscular tail and streamlined body, Otters are superbly adapted for a life in water. Although typically secretive, Otters have made a dramatic comeback and are increasingly spotted along our rivers.

- **Stoat** Short-legged and slender-bodied, Stoats are experts at hunting small mammals underground. In some populations their fur turns white in winter, when they become known as Ermine.

- **Weasel** This is a smaller version of the Stoat – indeed, it is the UK's smallest carnivore – with a comparatively stubby tail. Weasels are constantly on the move, and need to eat one-third of their body weight every day simply to stay alive.

- **Pine Marten** This cat-sized carnivore is a mustelid Jack of all trades, being as agile climbing trees as it is adept at swimming or hunting along the ground. Within the UK it is largely confined to Scotland, so seeing this chocolate-coated mammal makes for a red-letter day.

Above, top to bottom:
Otters are a primarily aquatic mustelid.

Weasels are a diminutive, short-tailed version of the stoat.

Pine Martens are the most arboreal of British mustelids.

Above: Polecats are spreading east from their core range in Wales.

Above: American Minks are much loathed by conservationists.

- **Polecat** This is a large, dark version of a Stoat, with a bandit mask across its eyes. Polecat populations are increasing and spreading from their stronghold in Wales. A major threat is hybridisation with escaped domestic ferrets, the offspring of which ('polecat-ferrets') are more abundant than their wild relative.

- **American Mink** As its name suggests, this voracious predator should not be in the UK at all. Imported to be bred commercially for its fur, deliberately released animals and escaped convicts have spread countrywide and wrought havoc on our native wildlife, notably the Water Vole.

European relatives

Five additional species of mustelid range across Europe. Two are Polecats –Steppe and Marbled – which resemble the species occurring in Britain. The Wolverine is a tough, heavily built carnivore inhabiting boreal forests from Scandinavia east. The European Mink (*right*) differs from its introduced American cousin by showing white on its upper lip; it has a similar diet, principally comprising aquatic rodents. The Beech Marten (*below*), also known as the Stone or House Marten, occurs widely in Europe. As its latter name suggests, it often frequents human habitation.

Above: European Mink is one of Europe's most endangered mammals.

Below: Beech Marten occurs widely across Europe.

Below: Wolverine is a fierce predator of boreal regions.

Becoming Badger

Above: Badgers exhibit typical physical characteristics of their family such as short legs and long spine. Their more powerful front legs, however, indicate that this is a burrowing creature.

It is 20 million years since the Badger's mustelid ancestors ceased climbing trees, and began to evolve traits that prepared them for a life on and under the ground. From then it took as many as 16 million years for a recognisable Badger-like form to emerge in the fossil record. Fossils identified as being of the same genus as today's Badger (*Meles*) are 2–3 million years old.

Modern Badgers retain their family's defining characteristics: long spine, short legs and enlarged scent glands on either side of the anus (the word 'mustelid' echoes the word 'musk'). However, Badgers differ by having developed physical features that serve for burrowing and digging rather than for climbing (as Pine Martens), chasing rodents underground (Stoats, Weasels) or swimming (Otters, American Mink). Among European mustelids (and, indeed, the European continent's mammals more widely), there is nothing quite like the Badger.

Badgers around the world

Clockwise from top left:
Large-toothed Ferret Badger is one of up to five members of the genus *Melogale*.

Up to three species of Hog Badger (*Arctonyx*) inhabit rainforests in south-east Asia.

The genus *Meles* contains the Badger(s) with which we are familiar.

American Badger is the sole representative of New World Badgers and the only member of its genus, *Taxidea*.

Like all scientists, biologists love arguing. The number of badger species, and the relationship between them, has been a subject of fervent debate among mammologists at conferences and in learned journals worldwide. These questions remain divisive, and contrasting views are likely to remain for decades.

At least the professionals can agree on some things. There is now a majority view that three so-called badgers from around the world are more closely related to other groups of mustelids, namely weasels, stoats and martens (in the case of the African Honey Badger *Mellivora capensis*), or skunks (two species of stink badger *Mydaeus* from Asia), than to badgers.

After this things get a bit murky. One prevailing view is that 'true' badgers are spread across just four genera.

First up, there are three to five species of ferret badger *Melogale* wandering through Asia. In that continent there are also between one and three species of hog badger *Arctonyx*. The New World has just a single species, housed in its own genus, the American Badger *Taxidea taxus*. This leaves the genus *Meles* for the Badger with which we are familiar.

The alternative attempt at a family tree places *Meles* rather differently. This splits the mustelids as a whole into eight subfamilies (a rather arbitrary taxonomic layer between family and genus). It argues that the ferret badgers and American Badger are each sufficiently distinct to merit placement in their own subfamily (*Helictidinae* and *Taxidiinae*, respectively). It groups together in a separate subfamily (Melinae) the *Meles* Badger and *Arctonyx* hog badgers.

If this were not bewildering enough, there is a further twist to meline classification. Until recently, *Meles* was thought to contain a single species: the Eurasian Badger *Meles meles*. But some mammal experts now think this is better split into a trio of separate species: the European Badger, with which readers of this book will be most familiar, the Asian Badger *Meles leucurus* and the Japanese Badger *Meles anakuma*. This remains controversial, however, so this book adopts the traditional and cautious view that there is just one species of *Meles*, the Eurasian Badger – and typically refers to it simply as 'Badger'.

Some biologists consider there to be three species of Badger in the genus *Meles*: Asian Badger (**above**) and Japanese Badger (**below**) as well as the familiar European Badger (**left**). We adopt the conservative classification of there being just one species.

Badgers across Europe

Above: Adaptable to a range of climates, Badgers live as far south as Portugal and Israel, and as far north as Finland.

Below: Across Europe as a whole, Badger numbers are stable or possibly increasing.

Fossils attributable to what we now know as the Badger (albeit the now extinct subspecies *atavus*) date back 0.5–1.5 million years. Each subsequent ice age forced a retreat southwards into refugia such as Italy and Spain. Following the most recent ice age, 15,000 years ago, Badgers restarted their northwards push, finally reaching Scandinavia 8,000 years later. The species now ranges over much of Europe and continental Asia, from Portugal and Ireland in the west to Japan in the east, from Finland in the north to Israel in the south. The animals live from sea level up to 2,300m (7,545ft) altitude.

Between them, Sweden, the UK and the Republic of Ireland hold more than half of Europe's 1.5 million Badgers. The three countries all have average Badger densities above one adult per square kilometre; concentrations are lower (often much lower) everywhere else in Europe. Northern Ireland hosts the most densely packed Badgers (2.75 per square kilometre). In Poland and Estonia, Badger densities are 70 times lower, so that there are, on average, only four individuals to a hundred square kilometres.

Even if the Badger populations of some large European countries are not very tightly packed, some of these countries nevertheless have sizeable Badger populations. Germany, for example, has about 140,000 *Dachs* and France just over half that number of *blaireaux*. Respectively, these countries provide 9.5 per cent and 5.3 per cent of Europe's total. Across Europe as a whole, as far as researchers can piece together from a jigsaw of data, Badger numbers seem to be stable or increasing – so the species' situation is not of overall global conservation concern. Good news for Brock, *Dachs* and *blaireau*!

Badgers in the British Isles

Fossil evidence suggests that Badgers had reached the island of Britain by 500–750,000 years ago. Subsequent ice ages, however, forced them to retreat southwards and eastwards. When the glaciers finally melted 10,000 years ago, Badgers trundled back across the land bridge that then connected Britain to continental Europe and re-established themselves there.

Upon arrival in Britain, Badgers would have lived alongside species that we now associate with the tundra and taiga of extreme northern Europe, such as the Arctic Fox, Reindeer and Wolverine. How Badgers reached Ireland is unclear, and no theory is particularly plausible. They may have used a temporary land link from Scotland (if one existed). They may have been introduced by Bronze Age humans. Or they may have persisted in a relatively mild Ireland throughout the ice ages.

Nowadays, Badgers are widely distributed across the British Isles – but neither ubiquitously nor uniformly. For example, they occur only patchily in the Scottish Highlands north of Inverness, and are absent from all major islands bar Anglesey (Wales), Arran and Skye (Scotland), and Canvey, Sheppey and Wight (England). In the British Isles, Badgers prefer lowlands (being rare above 500m/1,640ft altitude), and geology that enables them to easily dig burrows that will stand the test of time: heavy clay or peaty soils tend to be shunned.

In Great Britain (that is, England, Scotland and Wales), Badgers are thought to number around 300,000. There are about 38,000 in Northern Ireland and 84,000 in the Republic of Ireland. In Great Britain, the population is as high as it has been since the 18th century, when persecution of Badgers as pests became ever more intense. The Badger population reached its nadir in the early 1900s. The tables only turned during the aftermath of the Second World War. As the number of gamekeepers dropped (the combined legacy of battlefield mortality and the break-up of country estates), so Badgers resurged. By 1963 there were 36,000 Badger groups estimated in the UK, which

Above: Badgers recolonised the British Isles 10,000 years ago, following the most recent ice age. The islands' current population is more than 400,000.

Below: In Great Britain, Badgers have long been persecuted as vermin and pursued for 'sport'.

Above: Badgers typically emerge from their sett as the sun is setting.

Above right: Many Badger territories include some element of woodland.

Below: Badgers have been the subject of much research, with a particularly long-running site at Wytham Woods (Oxfordshire).

Below right: Agricultural flatlands offer few foraging opportunities for Badgers, but there might still be a sett in the distant woodland.

rose further to 43,000 by 1985. Still the population grew. In the ten years to 1997, the UK's Badger population was thought to have increased by 75 per cent.

In England, Badgers are most common in the south-west, which accounts for perhaps one-third of the total population in Great Britain. Densities here are often as high as 20 adults per square kilometre (compared with an average of 1.39 across Great Britain), with more than 30 per square kilometre recorded at one well-studied and well-protected site in Oxfordshire, Wytham Woods. Badgers are rarest in the agricultural flatlands of East Anglia. Here Badger groups are very few and far between – but perhaps becoming less so with each passing year. In Northern Ireland, densities are roughly double those in Great Britain; those in the Republic of Ireland are slightly lower than those in Great Britain.

Why do Badger numbers vary from place to place?

The reasons for differences in Badger population size and density, whether across a continent or between counties, are complex. Everything hinges on the interplay of several variables: climate, food supply, terrain for digging burrows, the frequency of disease and safety from persecution. Let's explore three examples.

First, climate and food supply are often linked. For the last century, Badgers have been expanding their range northwards in Scandinavia. This is probably because climate change has resulted in longer summers, so that there is more time for cubs to feed before winter sets in. In turn, this means that more yearlings survive, enabling the population to expand and colonise new areas.

Second, the comparative rarity of Badgers in the Netherlands is due to the flatness of the landscape, which makes it hard for Badgers to dig setts without them risking being waterlogged. Third, in countries where Badgers are or have recently been persecuted, the piebald mustelid is unsurprisingly scarcer. In much of Europe, Red Fox dens were gassed to control rabies during the 1960s and '70s – and this had the consequence of killing many Badgers, given that these two species often share underground retreats. The United Kingdom and Republic of Ireland have never suffered from rabies, so Badgers have never endured this particular fate.

Above: Waterlogged setts are an occupational hazard for Badgers living near watercourses.

Below left: Longer summers – a result of climate change – may be helping Badgers spread north through Scandinavia.

Below: Programmes controlling rabies in Red Foxes in Europe inadvertently killed Badgers.

Body Beautiful

A Badger may not have the grace of an Otter or the slink of a Weasel. It may lack the shiny sleekness of a Pine Marten or the textile-industry quality of an American Mink's fur. But for the purposes for which a Badger needs it, its body is unarguably beautiful – and in the cut and thrust of life, that is what really matters.

Catch sight of a Badger trotting across a field or road, and the overall colour impression imprinted on your retina is of a peppery silver-grey. No prizes then for deducing the rationale for the Badger's colloquial English name of Grey. In a more prolonged observation, it becomes evident that the colour and pattern of a Badger's fur is rather complex. Each 'guard hair' – the outermost part of the coat – is white for much of its length, but has a broad blackish band just shy of its tip. Beneath this external protection, and usually only visible on a soaked Badger, shelters dense white underfur that serves as thermal underwear.

Given this understated overall colouration, the Badger's head pattern is shockingly bold. A snowy-white head and neck is offset by an elongated black bandit-mask. This stretches backwards from a lump of coal of a nose, passing

Opposite: The Badger's head pattern is shockingly bold and instantly recognisable.

Below: The peppery grey impression given by the Badger's flanks and back is actually a complex combination of black, white and grey on each individual hair.

Above: Examined carefully, the Badger's head is actually white with a bold, flaring black stripe running backwards through each eye to the ear.

through the eyes to the back of the head, daubing all but the tips of the ears on its way through. Face-on, the Badger is a humbug of a mammal, with a quintet of alternate black-and-white stripes sourced from its muzzle. Is there a more widely recognisable 'face' pattern in the entire animal kingdom? Little wonder that the logo of the UK Wildlife Trusts now adorns road signs across the countryside.

Evolutionary theory demands a purpose for the Badger's pattern. One possible explanation is that the stripes break up the animal's form in shafts of moonlight. A more widely accepted rationale, first proposed more than a century ago and supported by recent research, is that bold black-and-white patterns serve as a warning. In this case the Badger is alerting potential adversaries or predators to its willingness to defend itself with a powerful bite. In the case of other pied mustelids, the skunks, the warning relates to smelly spray.

Below: One possible explanation for the Badger's bold face pattern is that it serves as an easily visible warning to potential predators.

Anatomic steroids

In *Badgers*, a comprehensive review of what we know about these iconic mammals, biologist Tim Roper suggests that – as a gross generalisation – animals 'survive either by being quick and agile or by being robust'. He argues that the Badger's accumulation of physical characteristics, which are unusual within both the mustelid family and in carnivores more widely, 'places the mammal firmly' in the latter category. In its anatomy, the Badger is definitively towards the body-builder end of the mustelid spectrum.

Badgers are typically 65–80cm (25–30in) long and 10–12kg (22–26lb) in weight, with boars 5–10 per cent bigger and heavier than sows. Compared with most mustelids, Badgers are low-slung and stocky mammals, roughly as broad as they are tall. The tube-like body is compact, the legs short; both are powerful. The neck, shoulders and front legs are particularly muscular, enabling the Badger to move rocks more than twice its typical weight of 10kg (22lb) or so. The strong forelimbs, combined with long, gently curved claws, are perfect for digging: just what Badgers need to excavate for both food and home.

If you were ever to see a Badger skeleton, you would probably be surprised at the weight and solidity of the skull. You might also wonder about the raised ridge that crests the skull and remark on the protruding jawbone. Combined with strong neck muscles, these cranial features give the Badger a formidable bite.

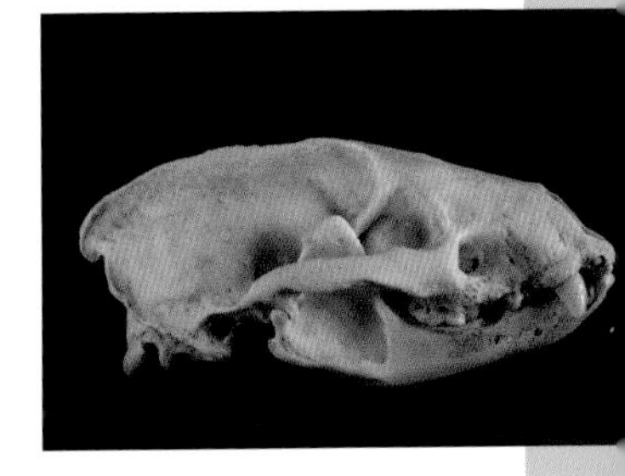

Above: A protruding jawbone and raised crest are part of the explanation for the Badger's powerful bite.

Below: Badgers are short legged and stocky, with a cylindrical body packed with muscle.

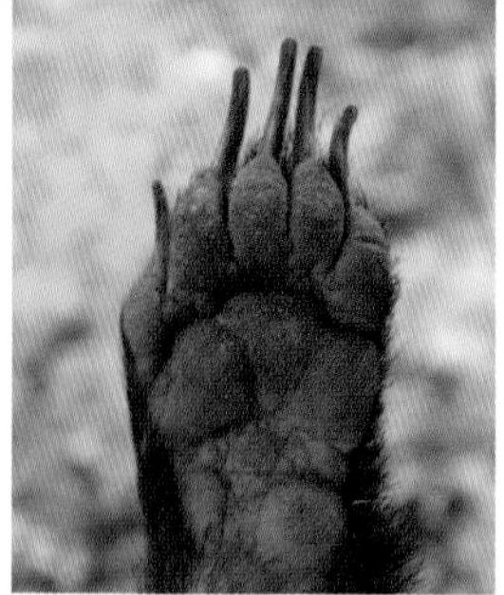

Above: Badgers' long curved claws are perfect for digging, but also serve for marking their territory by scratching on trees.

Below: Strong jaws and bite are thought to aid fighting, rather than serve in foraging.

Their diet may be cosmopolitan, but Badgers eat nothing that requires such power. This leads scientists to reckon that the bite has evolved as a means of self-defence against predators or rival Badgers. In combat the locked jaws enable a Badger to hang on to even a vigorously squirming adversary. As is the case with overall length and weight, the crest of a boar averages slightly taller than that of a sow. This suggests that the Badger's bite has evolved partly to help males fight over territory or mating rights.

Variations on a theme

Not all Badgers look the same. Some individuals have odd colouration, appearing tan, entirely black or wholly white. This is the result of a genetic quirk. If the aberrant individual breeds, the idiosyncracy may be passed to its offspring. As a result, colour variants cluster in certain areas. In Britain, for example, white individuals are comparatively common in Essex and Kent.

Across the Eurasian Badger's range as a whole, however, there are clear trends in pelage colour and head pattern. Badgers along the Mediterranean tend to be pale brown, whereas those in eastern Russia are very dark with brown replacing white on the head. Badgers in Asia (we have already learnt that some biologists argue that these are a separate species) are pale sandy-grey with markedly narrower black stripes on the head. Those in Japan (ditto) have an almost yellowish cast with the black on the head reduced to tear-stained mascara. This systematic variation lends support to those who argue that what we think of as 'the Badger' is actually three separate species. These are Badgers, but not necessarily as you know them.

Above: Genetic quirks can result in aberrant colouration – in this case tan – spreading locally through populations

Above: Badgers in eastern Russia can be very dark, and may represent a separate species.

Below: A typically coloured European Badger has a grey back and flanks with a paler tail, black-and-white striped head and black underparts.

It all makes sense

Above: The big eyes of a Slow Loris mark it out as a strictly nocturnal creature.

Below: With its small eyes and ears but large nose, it is clear that the Badger primarily uses its sense of smell to gain information on its surroundings.

Examining the Badger's external appearance enables us to have a good guess at the relative significance of its senses. Its eyes are small, suggesting that vision is not particularly important. While nocturnal mammals often have large eyes (think of a saucer-eyed Slow Loris or bushbaby), those that spend much or all of their lives underground have little need to see. The European Mole, for example, is almost entirely subterranean – and has eyes that are barely a single millimetre in diameter and, moreover, largely hidden or protected by fur. That a Badger's eyes are small rather than tiny is probably a trade-off between its sett-focused existence and the need to be able to spot movement or an unexpected silhouette on forays above ground.

The Badger's ears are towards the smaller end of the mustelid range, and equate to reasonable but not excellent hearing. The Badger's key aural objective appears to be to work out how far away a sound is coming from. Distant sounds, even if loud, can be safely ignored. However, a Badger will be on alert if it hears an unexpected noise at close range. This may explain why a Badger – unfortunately in terms of its survival – is more likely to start at the click of a camera shutter in a quiet woodland than at the roar of an all too rapidly approaching vehicle as the animal crosses the road.

In comparison to head size, no European mustelid has as large a nose as the Badger. Inside its spacious nasal cavity, numerous small, heavily pleated bones provide a large surface area for detecting scents. Because of this combination of external and internal adaptations, smell is the Badger's most important source of information. Upon emerging from its sett, a Badger's typical first act is to elongate its body, tilt its snout upwards and sniff the air. A refined ability to smell is critical for Badgers. Through smell, they find food, follow paths, detect danger, determine mating opportunities and differentiate friend from foe. From a would-be Badger watcher's perspective, this explains why it is essential to be downwind of a sett.

Do the locomotion

Above: Muscly legs mean the Badger is capable of a surprising turn of speed.

Badgers typically move delicately and methodically. The normal pace is a meandering, rather mechanical walk on stumpy legs. When feeding, a Badger's head wags constantly from side to side as it sniffs out its next meal. Badgers may trot rather than trundle when moving purposefully to feeding grounds at the start of the evening and between food-rich patches during the night. If it has sensed danger in particular, a Badger will break into a canter and, for short bursts, may accelerate into a surprisingly effortless gallop. A typical trotting speed is 7km (about 4½ miles) per hour, increasing to perhaps four times that when scudding along. Watch a running Badger and you will recognise the rippling gait that characterises all mustelids, from Stoats to skunks.

Badgers further betray their mustelid affinities through two other modes of movement. We may think of them as creatures confined to life on or below ground, but Badgers are able to both swim and climb. Granted, Badgers are not Otters when it comes to water, but they can propel themselves proficiently with a doggy paddle, head held high. In a similar vein, a Pine Marten would shin up a trunk far quicker than a Badger, but the latter is entirely able to ascend small trees to eat slugs, clamber over fences and walls, and pull itself up onto troughs to munch on cattle feed.

Above: Long spine and short legs lend Badgers the rippling gait characteristic of their mustelid family.

Below: Although primarily ground-living creatures, beetling about for grubs atop a fallen trunk poses a Badger no barriers.

An atypical mustelid

Above: Badgers are proficient climbers, readily ascending large trees or cattle troughs to reach food.

Above: They may not be as at home in water as Otters, but Badgers can certainly swim when needs must.

While its movements may betray the Badger as a mustelid, in several ways it is the odd one out in the family. Whereas all other British mustelids are uniformly active year round, Badgers spend the winter largely or wholly inactive in 'torpor'. Indeed, according to a single study, Badgers may have a markedly lower metabolic rate than fellow mustelids, its level being closer to that of exclusively burrowing mammals.

The Badger is not the only European mustelid – a carnivorous family, remember – to be omnivorous. However, while the Pine Marten and Stoat, for example, seasonally supplement their meat-rich diets with berries, both fall way short of the Badger's routine omnivory. This difference reveals itself in the animals' teeth. Mustelids typically have the sharp dentures of a carnivore, but the Badger's dentition is more ambivalent, exhibiting the broad, flat molars associated with omnivores.

Finally, the Badger is a tank in mustelid terms: large and stocky, muscular and thickset. Whereas Stoats and Otters escape danger through agility and speed, the Badger's survival strategy focuses arguably on hiding (hence setts) and certainly on fighting (hence those neck muscles and jaws). An atypical mustelid, indeed.

Right: Unusually amongst British members of its family, the Badger typically sleeps for much of the winter.

Right: Mustelids are a carnivorous family, but the Badger is routinely omnivore.

Below: If Stoats are built for speed, Badgers are a tank – and not to be messed with.

Breeding and Growth

Scientists have learnt a lot about how Badgers breed – but admit that there is more they have yet to fathom. Most Badgers reach sexual maturity shortly after their first birthday. This is an average, of course. In mild regions some precocious females are able to conceive earlier than this, when just nine months old – but they seldom breed successfully. In contrast, some late-developing males are not equipped to mate until they enter their third year. In countries with harsh winters, maturity is set back a year as it can take two full summers for youngsters to reach adult weight.

The peak age for successful breeding in both males and females is typically between four and six years old. Most adult members of a social group mate at some point during the year. However, less than a third of adults actually produce cubs in any one season. Moreover, typically only the group's dominant female breeds successfully. This suggests that something strange is going on…

Opposite: Badger cubs take time to gain the confidence to explore above ground without their mother.

Below: Badgers can reach sexual maturity in their first year, but most successful breeders are between four and six years old.

Courtship

At the risk of entering the domain occupied by racy novels, the act of mating itself is fascinating. The boar initiates the process, wooing the sow through a courtship display. He 'churrs' to her, raises his tail to 90 degrees and struts around on stiff legs. Sometimes the male complements this by squatting on the ground to leave a scented message that invites the female to emerge from the sett (romantic, eh?). Even if the sow accepts the invitation, there is no guarantee of mating. If the sow is not 'in the mood', she will reject the boar's advances and may even flee the scene.

If the duo grooms each other or frots one another with the scent glands on their backsides, however, the omens are good. When it happens, mating lasts between a few minutes and 90. Only longer, stamina-sapping copulations appear to result in conception. The purpose of shorter, unsuccessful matings may simply be to prompt the female to ovulate. (With apologies if this causes you to blush, there is even a school of thought that the male's penis bone accelerates ovulation.) Once mating is over, the sow usually retires to the sett, leaving the boar to head out into the night.

Below: Badgers let it be known that they are 'in the mood' by leaving scented messages for suitors to find.

Above: In Northern and Western Europe, the principal Badger mating season is late winter.

There is a clear seasonality to Badger procreation. Females typically ovulate between January and March in the UK, but a month later in Sweden. Based on information from France and Scandinavia, male testes swell to their largest size in February and March, lending weight to the notion that the large majority of matings takes place in those two months. However, some males appear to be fertile year round, and most females ovulate pretty much each month until they become pregnant (with particular peaks in January–March and July–September).

Delaying the moment of truth

By the end of April eight adult females in ten carry fertilised eggs. They are not, however, actually pregnant. Sows delay implantation of the egg until December. Amazing though this may seem to our own species, Badgers are not unusual in having such a strategy. Indeed, it is so common among mustelids that half of all mammal species known to delay implantation are relatives of the Badger.

The strategy gives cubs the best start in life and strengthens prospects of them surviving their vulnerable first winter. It's a cunning ruse that times birth for late winter, which subsequently means their emergence above ground coincides with spring's increase in food availability. Given that pregnancy lasts only seven weeks, due to delayed implantation Badgers do not need to use up valuable energy mating during winter, when they are dormant.

Delayed implantation offers sows two additional benefits. First, over summer and autumn they can ovulate again, and mate with the same or a different male, and thus top up their carton of fertilised eggs. This is particularly useful for females that lost eggs after the spring mating, as is the case for up to 60 per cent of sows. Second, with implantation taking place as late as December, just before the retreat underground for winter, sows may make the most of autumn's glut of food, storing up fat for eventual conversion into milk to nourish the cubs.

Who's the daddy?

The late-winter peak in matings occurs shortly after sows have given birth. This initial window of opportunity is short: sows are only receptive for five or so days. If boars wish to be sure of their paternity, this is an important time for them to 'guard' their sows. This is because males and females have different motives. In simple terms, males want as many descendants as possible, whereas females want the best-quality male genes to give their offspring the optimum chance of survival.

Sows often hedge their bets, straying into neighbouring territories while fertile, and mating with different males. Studies have shown that Badger promiscuity is rife. Illicit

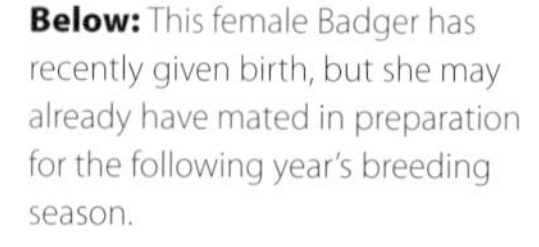

Below: This female Badger has recently given birth, but she may already have mated in preparation for the following year's breeding season.

Above: Badgers are so promiscuous that it is possible for these two cubs to each have a different father.

affairs result in half of all cubs born. Even more remarkably, cubs born in a single litter can have different fathers. Rarely has it been more pertinent to ask 'who's the daddy?'

The pregnancy equation

There is another question to ask, and one that has a somewhat insidious answer. In each Badger social group, 'who gets to be a mummy?' We know that eight females in every ten carry fertilised eggs by the end of April. Yet normally only one female in every group succeeds in raising cubs. Indeed, of every 100 fertilised eggs being held by females, just five cubs emerge above ground in spring. This is such a poor success rate that something odd must be going on.

The main reason seems to be strategic and rather clever. Whether a female will raise cubs seems to come down to two factors: her age compared with the ages of others in the group, and her physical condition. Much can change in a Badger's year: summer could be a season of plenty (or not); a car could kill a dominant female (or not). This means that females have no way of knowing how life will have panned out by December, when fertilised eggs are implanted and the journey to giving birth truly starts.

While pregnancy uses up lots of energy, and giving birth even more, little cost is attached to allowing eggs to float around in the womb. This means that females might as well mate early in the year, then 'take a decision' as late as December on whether to proceed to pregnancy.

Below: In terms of energy, breeding is a very costly process so female Badgers only proceed with pregnancy when optimistic of success.

The 'decision' is unlikely to be conscious, but rather an in-built, automatic biological 'get-out clause'. Somehow the sow weighs up the prospects of a successful pregnancy against the costs of getting there.

If the odds of a successful birth are high, the sow will go for it. If things are looking dicey – for example, if the harvest has been meagre and the sow is scrawny – then there is no benefit, and much cost in terms of energy, in continuing to pregnancy. The sow is better off cutting her losses by ejecting or reabsorbing the eggs, and trying again the following year.

All this makes evolutionary sense. But there's a twist. There is another reason why only a tiny proportion of fertilised eggs produce healthy cubs. This time the evolutionary driver is rather more unsavoury. Each female stands the best chance of her cubs surviving their first winter if there is less competition from other females with cubs. The way the dominant sow reduces that risk is brutal. She harasses and fights potential rivals so that their eggs fail to implant or, if that stage is passed, they miscarry.

In the UK at least, sows that succeed in remaining pregnant retreat underground during December. Huddling with fellow group members to conserve heat, they also conserve energy by reducing activity levels to the bare minimum. All being well, the sows give birth seven weeks later.

Below: Badgers typically give birth to litters of up to five cubs.

Cubs' honour

Cubs are born underground in the warmth and seclusion of the Badger group's main sett. On average, UK births peak in the first fortnight of February, with three-quarters happening in the two months from mid-January. Unsurprisingly, births are earlier in southern countries where winters are milder (for example, early January in southern Spain and late January in south-west France), and later in northern countries where snow may lie on the ground well into spring (for instance, early March in Sweden).

Litter size varies from one to five cubs; the average is nearer three than two. Cubs are born into the lightless world of the burrow both helpless and blind. Their loose pink skin is sparsely covered in curt silver fur. At 12cm (4¾in) long and 90g (3oz) in weight, newborns are about one-sixth the length of their mother and 1–2 per cent of her weight.

Over the next four weeks, each cub's coat gradually becomes a faded version of an adult's, and its mother's milk enables rapid growth. In the following two weeks, the cubs' milk teeth burst through and their eyes open. The nippers then spend a fortnight exploring their underground haven until their mother decrees that they have developed sufficiently to experience the big bad world above ground.

Unfortunately, not all cubs get that far. Around one-fifth of youngsters never make it outside, dying in the sett.

Below: Newborn Badgers (left) are helpless and blind, but within four weeks (right) have developed the familiar black, white and grey pelage.

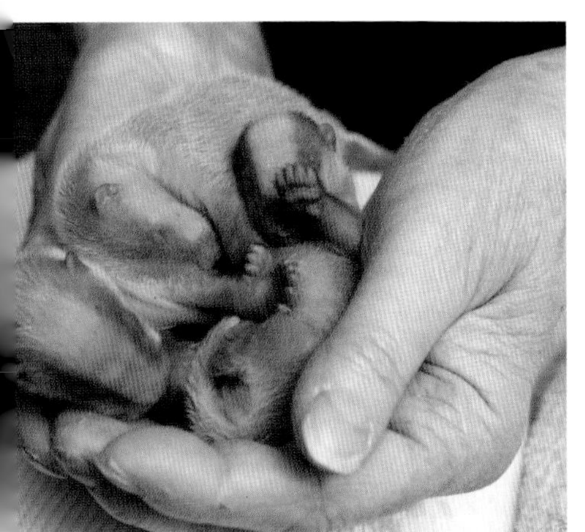

Above: The sett is an underground haven where Badger cubs grow, sleep and play for the first eight weeks of their life.

Another fifth succumb before they have been weaned. Rather than each female losing one or two cubs on average, this mortality is largely due to a proportion of females losing their entire litter. In both cases, a mother's inadequate milk supply may be the reason. However, as is the case with pregnancy, there also appears to be an unpleasant explanation for the high mortality rate, namely infanticide. There is some evidence to suggest that dominant females sometimes kill the cubs of another female in the group, presumably to minimise competition for food resources with their own offspring.

Below: Badger cubs are usually two months old before their mother allows them above ground.

Expanding horizons

At first cubs venture out only late at night when, their mother apart, the group's adults are foraging widely. By 12 weeks old the litter emerges at the same time as the adults, shortly after dusk. Cubs spend longer periods above ground and start to range further from the sett. Within a month they have adult teeth, are doing whatever adults do (except mating, of course!), and play vigorously among themselves and with other group members. This fourth month of life is an intense period of learning, relationship building and social integration.

Cubs forage with their mother throughout the summer and autumn. Nature's ample bounty enables the youngsters to grow rapidly. They reach adult size within six months of birth, and adult weight by the time winter sets in. Building up fat reserves is critical to cubs surviving their first winter. One estimate suggests that fewer than half the cubs emerging from the sett survive to the following spring.

In most mammals, males leave their parents' territory once they become sexually mature. For them, striking out for pastures new is necessary to find a mate, and – in some cases – to avoid the wrath, or worse, of their fathers should they stay. In Badgers, things are rather different.

Above: Aged three or four months, Badger cubs gradually explore further afield from the sett.

Below: Play is a key part of a young Badger's development, enabling it to build relationships.

Above: Most Badger populations are gregarious, and there is much movement between social groups.

Should I stay or should I go?

Below: Badgers communicate through scent, and often defecate at the border between territories to leave messages for neighbouring groups.

Many individual Badgers stay with the same group for their entire lives. Those that leave may be of either sex and any age. They usually only go as far as the neighbouring group, where the process of integration takes as long as nine months. Some adventurous animals disperse much further. Adventure can come at great cost. The wider Badgers wander in search of pastures new, the greater the likelihood of running the gauntlet of road traffic.

Deciding where to strike the balance between staying and going is tricky. Individuals that stay put are betting on rising through the group's hierarchy to become the dominant boar or sow. Animals that depart are banking on doing the same within a new group, perhaps obtaining sneaky copulations en route, or fighting their way to the top.

One way for Badgers to inform their decision is to find out what is happening in a neighbour's territory, specifically whether or not there is a breeding vacancy. This can be discovered by popping over there to have a look around or – less riskily – by checking the 'scented messages' left at latrines along the territory border.

Above: These siblings may only be play-fighting, but the later rivalry may be for real.

Reducing in-breeding

In-breeding is where family members breed with one another. It is not a good thing as a too closely related family can increase the chances of genetic problems. Scientists used to think that Badger social groups were so rigid and self-contained that in-breeding was inevitable. We now know differently.

Badgers avoid in-breeding in three ways, two of which have already been mentioned. The most important mechanism is illicit matings with neighbours, and another is dispersal. Additionally, when mating within the group Badgers tend to do so only with a partner that has arrived from elsewhere. This reduces the chance of them being related. A measure of how successful Badgers are at avoiding incest is that England's whole population is genetically pretty similar. This can only be because groups have bred with their neighbours, who have bred with their neighbours, and so on.

Above: Badgers use various mechanisms to avoid in-breeding.

Below: Britain's Badger population is genetically similar, which suggests that adjacent groups routinely interbreed.

Social Life and Communication

As recently as the 1970s, scientists knew that Badgers lived in family groups, but little more. They didn't understand how these family units were organised or what purpose they served. Biologists had no inkling as to how far Badgers ranged when they left their sett, or whether there were any limits to those movements. They suspected that Badgers were strong, silent types, short on the desire to communicate.

The research of Hans Kruuk at Wytham Woods, near Oxford (UK), changed these perceptions. Indeed, Kruuk's insights and revelations opened the floodgates of Badger research. As a consequence, we are now considerably clearer about how Badgers organise themselves and to what end (their social life and social structure), how territorial they are and why, and how they communicate and what they feel the need to say. This firmer understanding suggests that Badgers are rather unusual among carnivorous mammals.

Most carnivores live on their own, coming together only when hormones decree it to be mating season: think of Tigers, Wolverines and Stoats. Fewer than one carnivore species in six lives in social groups, typically cooperating

Opposite: The vast majority of carnivores pursue solitary lives, only coming together to breed; the Badger is one of the exceptions.

Below: In Western Europe, Badger groups typically comprise between four and eight adults.

Above: In some parts of Europe, Badgers live in what we might term a 'nuclear family', comprising a pair and its cubs.

in order to thrive, for example by hunting in packs: Lions, Grey Wolves and Giant Otters are good examples. A tiny proportion of carnivores lives in social groups, but without getting any obvious benefit from doing so. Guess what camp Badgers fall into? Correct! Badgers are 'social carnivores' that apparently live *alongside* one another rather than *with* each other.

A 'social carnivore'

In most of Western Europe, Badger groups average between four and eight adults, with the ends of the spectrum set at two and 23. Typically sized groups occupy a territory of 20–100ha (50–250 acres), the top end of that range being equivalent to a square kilometre. However, there are notable exceptions. In Spain group size is smaller and territories are larger. In Switzerland and Poland the typical group comprises what we humans would call a 'nuclear family', with single adult male and female, plus their cubs, and the territory can be as big as 1,000ha (2,470 acres). Japanese Badgers – potentially a separate species, of course – may even live alone.

So, in places that cannot support many Badgers, group sizes are usually small but territories are large. Scientists have yet to agree *why* this is. The main factor may be food availability. Larger groups tend to be in areas with

Below: The size of a Badger group's territory may be linked to the availability of food.

Above: Areas containing a high density of lobworms have been found to support a high density of Badgers.

abundant lobworms, for example. More Badgers cramming into food-rich areas would reduce territory size and increase population density. Whatever the reason, unlike its mustelid relatives, the Badger is definitively a social animal. What is surprising (and confusing) is that it does not appear to be a *sociable* one.

We're gonna live together

In the 1970s, Hans Kruuk was mystified by the Badger's take on sociality. Here was an animal that lived in groups, but shrugged away the benefits of communal living. Badgers appeared uncommunicative, foraged alone and ignored one another when their paths crossed above ground. Bewilderingly, Badgers even shunned cooperative behaviour such as helping to raise the group's offspring, alerting one another to danger, fighting off enemies and spilling the beans about the location of food.

Below: Whilst Badgers are social mammals, they are not particularly sociable, shunning cooperative behaviour and often ignoring one another.

Above: What chance these two Badgers are related? On average, British Badgers are less related than a grandchild is to its grandparent.

In classical 'cooperative breeding systems', as biologists call them, helpers care for offspring of other group members. The theory is that helpers might be siblings of breeding animals, so pitching in strengthens the chances that their own genes will be passed on. Flipping this theory onto its back, one possible reason for Badgers' lack of collaboration is that members of a group are *not* closely related.

In Britain at least, reality does indeed roughly match theory. Due to regular movement between Badger social groups, on average, group members are less related than a grandchild might be to its grandmother. Of course, the downside of this idea being right is that we still don't entirely understand why Badgers bother living together.

One plausible explanation comes down to saving energy. Across most of their range Badgers are relatively or entirely inactive during winter, largely staying tucked up in their setts and lowering their metabolic rate. In countries with lengthy winters, Badgers stay in torpor for longer. This enables them to live off their accumulated fat layer for longer. Another good way to slow the rate of fat burn in winter is to snuggle together, keeping warm in a cosy underground chamber. Support for this idea comes from group members typically slumbering together in the main sett during winter and early spring, but splitting up and sleeping apart during the milder months of summer and autumn.

Below: The main attraction for Badgers in social living may well be to retain heat and reduce energy expenditure by snuggling together in winter.

Badgers as landowners

If a Badger's sett is its castle, then its territory is the accompanying country estate. Just as it costs more money for landed gentry to erect high boundary walls around bigger properties, so it costs Badgers significantly in energy terms to defend a large territory. For Badgers with small territories, however, marking the extent of the group's dominion is a relatively minor job, particularly when it is shared out among its members. Moreover, it is an important job because, as we have seen, small territories go hand in hand with high population density. This means that there are plenty of neighbours coveting adjacent property – so ample potential for a land grab.

All this results in Badger territoriality being more frequent in areas of high population density. The main way that Badgers communicate the extent of their territories is through odour. This includes spraying with scent from the anal glands, but also through wee and poo. In high-density populations, Badger loos ('latrines' in scientist speak) are spaced regularly along the well-trodden territory boundaries and around the main sett.

What exactly are Badgers defending in their territory? A logical answer would be food – logical but, according to research carried out by scientist Tim Roper in Sussex (UK), untrue. Roper discovered that individual Badgers

Above and below: Setts form the centrepiece of a Badger group's territory, whether in open country (above) or woodland (below).

Below: A large expanse of bare earth on a woodland slope is a sure sign of a Badger sett.

Above: Badger latrines are typically found around the sett and along territory boundaries.

regularly snuffled edibles inside an adjacent territory. Such trespassing – unchallenged by the landowners – turned out to be particularly frequent in summer, with a few autumn incursions but almost none in spring. Roper also learnt that latrines along territory borders were mainly used in spring, with a lower peak in autumn. On the assumption that Badgers needed to wee and poo at least as much in summer as in those two seasons, Roper considered this rather odd.

The link between libido and land

Putting these two findings together, Roper concluded that Badgers are only territorial on a seasonal basis. In summer Badgers are entirely relaxed about trespassers. Given that summer is when Badgers forage the most, Roper reckoned that territoriality must not be about defending food resources such as earthworm patches. He worked out that latrine use peaks yet the frequency of intruders crashes during spring, the very season when most cubs are conceived. This led him to suggest that territorial behaviour is linked to mating, with boars defending 'their' sows from the libidinal attentions of interlopers.

This line of thought may also shed further light on why Badgers live together, which – as we have seen – scientists have yet to convincingly explain. Most mating happens underground in the sett. This suggests that living together in the same sett, at least during late winter and spring when mating peaks, enables boars to prevent 'their' sow from straying. Ultimately, everything comes down to sex!

Below: Badgers are most intolerant of neighbours during the mating season, and trespassing may result in fights.

Underground society

Above: In Badgers the sex ratio is only even among cubs; in adults, females substantially outnumber males.

In any one Badger population, at any one time, the sex ratio is uneven. The starting point is the same: equal numbers of male and female cubs are born. However, in adults there are typically about three females to every two males because males typically die earlier than females. The higher mortality rate of boars is probably due to a combination of factors, such as being weakened by territorial fights and being hit by traffic when dispersing in search of new territories or surreptitious matings with female neighbours.

Most social mammals (and many non-social ones) have a well-established system of dominance, where one individual is 'subordinate' to another, taking its place behind the more dominant animal in the queue for food, mating and so on. Typically, males are higher in the pecking order than females, and large individuals are dominant over small ones and older animals over youngsters. This is the dominance system displayed by Badgers.

The concept of a pecking order is particularly key for females. All sows able to breed should be trying to do so each year, but normally only one from each social group breeds successfully. This suggests that females compete for the right to breed – but that only the dominant female succeeds. Indeed, the matriarch sits atop the group hierarchy, with all other members subordinate to her. The key advantage of her position is the right to procreate.

Fight club

In any mammal social system where some individuals are dominant over others, conflict is inevitable. Merely sizing each other up and flaunting muscles cannot hope to resolve every position in a group's hierarchy. As fighting is costly for the victor as well as the vanquished, physical battles are usually the last resort. This is true for Badgers, particularly during the breeding season. At some point in their lives, most Badgers end up in a noisy, lengthy and bloodstained scrap.

Below: Fighting is costly for both parties, and generally only pursued should other means of conflict resolution prove unsuccessful.

Below: A classic bite-wound inflicted by one male Badger on another.

Chasing each other, Badgers attack with their teeth, putting sharp canines and sinewy neck muscles to optimum use. Typically, fighting occurs between two males or two females; cross-gender combat is exceptionally rare. Boars are more prone to fighting than sows. Males of neighbouring groups tussle for access to a fertile female. Females skirmish within their own group to prevent subordinate sows from becoming pregnant or rearing cubs. In these duels females tend to bite one another's faces and necks, whereas males go for the rump. Bloodshed is common, death not infrequent. Such outcomes reflect the high stakes: the right to breed.

How Badgers communicate

At the start of this chapter we learnt that scientists used to think of Badgers as 'strong, silent types' with little interest in communication. What we now know suggests that there is *some* truth in this, but that Badgers share information more readily and in more varied ways than we could ever have imagined.

First, the element of truth. When Badgers emerge from the sett in an evening, you might observe a bit of mutual grooming or, in the right season, play between cubs. However, it doesn't take long for Badgers to head out into the night on their lonesome, ploughing separate furrows. Group members typically stay well clear of each other when foraging. Even if animals end up feeding in the same field, they snuffle and snaffle alongside (or even despite) each other rather than together. Even more surprisingly, if one animal senses danger it looks after itself (or, if it is a nursing sow, its cubs as well), but doesn't bother warning fellow group members. This is very surprising for mammals that live in communities: just think of Meerkats.

However, to consider a Badger as uncommunicative would be way off the mark. The more we learn about Badger confab, the more sophisticated Badgers' messaging seems to be – and the more we concede remains to be deciphered. Let's take each mode of communication in turn.

Above: An alert Badger may have smelt danger.

Above: Badgers largely communicate through scent.

Below: This Badger may be preening or could be adopting a submissive position in the presence of a dominant adult.

Visual and vocal

Biologists have deciphered a dozen or so visual signals – postures, movements and more. These seem to be used in times of fear, defence, aggression and mating. Male Badgers sometimes stand on their hind legs, presumably to look as big as possible. Bears and Otters do the same thing, for the same reason. A Badger ripples its body when threatening another creature. Crouching low to the ground with facial lines flaring signals an intention to attack.

A scared Badger may fluff itself out by raising the hairs on its body, making it look rather like an enormous spherical hedgehog or even a porcupine. When simply terrified or to indicate submissiveness, as a youngster might be when confronted by the dominant sow, a Badger makes itself as small as possible and hides its face mask. Finally, before or during mating, boars make a couple of movements that presumably signal dominance: strutting on stiff legs and raking their rear paws on the ground. Quite the poseur…

For a supposedly taciturn animal, the Badger turns out to make a veritable orchestra of sounds. At least 16 different noises have been defined by Josephine Wong of Oxford University, each serving one of five purposes.

To keep in contact or reassure one another, adults and cubs both grunt, mothers purr to their cubs, and cubs respond by chirping, clucking or cooing. If a Badger is startled, scared or hurt, it may snort, yelp, squeak or wail. An angry or aggressive Badger barks, hisses, growls, snarls, and 'keckers' or 'whickers' like horses in a stable. Two sounds are associated with mating. If a sow feels harassed by a boar, she may 'chitter' as a way of telling him to back off. When a male is trying to coax a female to mate, he 'churrs' to her. The two sounds may often be sequential… Finally, a variety of noises is associated with playing animals.

Scented messages

As already noted, Badgers are closely related to skunks. They also have a powerful sense of smell, and deploy this to great effect when foraging. Put these two things together and it is little surprise that much Badger communication involves scent.

Below: Badgers use their acute sense of smell to forage, communicate and check their environment for danger.

Badgers produce scent in several ways. They release a musky odour both from inside their bottom ('anal glands') and below their tail ('caudal gland'). Intestines have glands that add smell to poo and, unsurprisingly, Badger wee also comes with its own particular perfume. These scents contain information of interest to Badgers inside their social group and beyond. The key messages relate to group identity ('you're in my gang'), territoriality ('get orf moi land', or 'she's mine, not yours'), or mating ('I'm in the mood').

Badgers leave scented messages using two systems: they squat to 'mark' *places* (normally the ground) and other *individuals* (called 'allo-marking'). When a Badger

'squat-marks' it lifts its tail, briefly rubs its bottom on the ground and smears it with a thick liquid from the glands inside or outside its bum. All Badgers squat-mark, but reproductively active individuals do it particularly often – presumably to tell the Badger world that they are ready to breed.

Allo-marking takes two forms; neither is presumably as unpleasant for a Badger as it would be to us. Either one Badger does it to another, or two Badgers do it to each other. In the first instance it is mostly done by males in winter or spring. Mammologists have yet to work out *why*, but – given the timing – it is likely to be something to do with breeding status. Perhaps the boar tells females that he is ready to mate and warns other males not to bother trying. The second method serves to share bacteria between group members, giving them a distinctive and unique 'group smell'. This strengthens the gang's shared identity, like tattoos do for the Russian mafia.

Clockwise from top left:

Being low slung helps a Badger squat mark.

Badgers release scent from within their bottom and below their tail.

Smelling the ground enables a Badger to determine whether a friend or foe passed by.

Members of a Badger social group have a shared smell that enables them to differentiate ally from rival.

Above: Whilst all mustelids have anal glands, only Badgers complement these with caudal glands either side of their tail.

Starting at the bottom, literally, let's look in a bit more detail at how these scents are produced, where they are placed and what they say. The chemical make-up of secretions from glands inside the bum is remarkably consistent between individual Badgers, regardless of age, sex or status. This implies that the fatty fluid conveys the message that 'Badgers are here'. However, the peak periods for Badgers leaving these liquid notes at latrines coincides with the main months for mating. This suggests that the purpose is also to advertise a willingness to mate (by sows), or to deter rivals from chancing their arm (males).

All mustelids have anal glands, but only Badgers have caudal glands. And what glands these are – no mammal has bigger ones. The secretion released from the caudal glands differs chemically from individual to individual. Interestingly, however, the make-up of secretions produced by members of the same group is relatively similar. This 'shared smell' helps strengthen the band's brand: 'you smell like me, so you must be a friend or relative'. As Badgers from neighbouring groups have different smells, it also enables detection of outsiders: 'you smell funny, so you may be a foe'.

There may be even more to this scent than meets the nose. Male caudal glands are double the size of females', and the build-up of the soapy-smelling substance is highest in winter and lowest in autumn. This suggests that, among males at least, there may be some form of mating or territorial function – this is yet another secret for biologists to unravel.

Toilet humour

The 'yuk' factor accelerates through the Badger's addition of scented notes to its poo and wee. Clearly, excreting waste material is a necessary law of life. Some of what goes in must come out. Yet only half of Badger visits to latrines are to do their business; the other half are to read the runes or to post their own status updates.

If latrines – which comprise collections of shallow hollows called dung pits – serve as message boards, they need to be sited where fellow Badgers will find them. Many latrines lie beneath trees, perhaps to protect the

Below: Badgers use latrines to 'do their business' and to share various signals.

contents (and their 'press release') from being washed away by rain. Most latrines are located either close to the main sett, along the territory boundary or where Badger paths cross. Indeed, two-thirds are along borders. This suggests that one purpose of poo clues is to emphasise ownership of the territory. Strikingly, however, neighbouring Badgers *share* latrines along territory borders. This suggests that Badgers use latrines to exchange information about who is in their gang, rather than to simply erect a 'no entry' sign.

More pits are dug and used in spring and autumn than in other seasons. Given that these are also peak periods for mating, scientists deduce that, in addition to defining a territory, the contents signal breeding status. For a female this might be a covert invitation to males from neighbouring territories to pay her an illicit visit. For males it is a warning to rival boars to resist the temptation to do so – or else – and perhaps also a sly invitation to sows to nip across the boundary.

This works as an explanation for latrines along territory boundaries, but what about dung pits near the main sett? Given that intruders are unlikely to venture this far into another group's home range without being challenged, messages left at the sett latrine must be internal memos directed at fellow group members. Most likely, individuals are reminding others of their status in the group's hierarchy.

Badgers poo solely in latrines, but pee there just half the time. The rest of the time they urinate elsewhere. Badgers wee on the ground, even when walking or squat-marking, and wee on the ground around a sett. This may not seem surprising; after all, we humans wee more than we poo. However, there is much more going on with Badger wee.

Scientists know this for several reasons. Badgers urinate in latrines most frequently in spring, and males use them more often than females. This suggests that urine communicates messages about mating rights ('she's mine; stay away'). Badgers also pee at prominent points on paths, such as where the mustelid thoroughfare meets a fence. This suggests a role in marking the territory, but scientists have not yet unravelled what exactly Badgers are saying. As is a common refrain throughout this book, we have so much still to learn about Badgers.

Examining Badger poo reveals what they have been eating: cereals (**above**) and earthworms (**below**).

Food, Home and Habits

Badgers can make a home pretty much anywhere as long as the climate is amenable, there is adequate food and the terrain is suitable for digging a sett. But what, for a Badger, constitutes amenable, adequate and suitable?

In the UK Badgers typically inhabit countryside with a mixture of land uses, but some have adapted to live in urban areas. A prime territory would fuse pasture and arable land adjacent to deciduous woodland. Collectively, the three areas provide food throughout the year; as we shall see, Badgers are highly versatile omnivores with a dietary range wider than that of any other British mammal. Providing it was on top of sloping ground and well-drained soil that could be easily dug, the woodland – or some other sheltered vegetation such as a hedgerow – would also provide protection for the sett.

Until recently scientists assumed that such habitat requirements applied to Badgers throughout their range. They realised their mistake once results of research in continental Europe were published. These revealed the Badger's remarkable adaptability. Badgers, it transpires, live in places as diverse as Spain's coastal sand dunes, Switzerland's Alpine forests and intensive arable plains in Belgium, not merely in deciduous woodland next to fields grazed by cows.

Opposite: Badgers use a wide variety of habitats, but British animals typically include deciduous woodland as part of their territory.

Below: A male Badger emerges at the start of its nocturnal foray.

Above: In Britain at least, Badger setts are typically on sloping ground and under wooded cover; extensive bare ground often makes spotting them easy.

Below: An adaptable creature, Badgers inhabit terrain as varied as Coto Doñana in Andalucian Spain and Lake Silvaplana in the Swiss Jura.

In central Spain, Badgers appear to prefer comparatively hilly terrain at mid-elevations. Here they favour pine forests and *dehesas*, open woods with pastures. If there are watercourses and rocky areas within such habitats, all the better as far as the Spanish Badger is concerned. In Switzerland, specifically in the Jura Mountains, Badgers switch habitat allegiance with the season. During winter and spring they use forests and wooded pastures. Come summer and autumn, however, they move to grain fields.

City slickers

Badgers are an increasingly common sight in urban areas. In the UK, they occur in several areas of London (**above**) including within the confines of the Royal Botanic Gardens, Kew (**below**).

As far as rural Badgers are concerned, this is all well and good, but what about their bands of urban brothers? Unlike Foxes, Badgers may not be as routinely associated with towns and cities as, for example, Red Foxes in Britain. That said, published reports provide evidence of Badgers across eight countries stretching from the Republic of Ireland to Japan, and including at least 11 towns in the UK. This suggests that there may be many more Badgers than we might think in built-up areas. Many live within spitting distance of our homes and gardens – and their number seems to be on the up.

In Britain, the surge of Badgers into the 'burbs seems to have happened since the 1960s, around the time when populations overall were gathering a head of steam. British Badgers now inhabit towns from Bristol to Birmingham, Edinburgh to Exeter, and Nottingham to Newhaven. Setts can be spotted in secluded parks and cemeteries, allotments and wasteland. In London, there are even a score of Badgers in the world-famous Kew Gardens.

Above: Britain's urban Badgers have markedly smaller territories than their rural brethren, as they have learned to scavenge up to half their food intake from our discards.

'Townies' differ from their rural brethren in several ways. In Britain group structures are not so fixed, with a higher proportion of Badgers living alone or 'floating' between groups. Territories are markedly smaller than in the surrounding countryside – indeed, the smallest recorded anywhere, leading to the highest density of adults ever discovered (33 per square kilometre).

The reason for urban Britain's smaller territories is food availability. In cities, there is so much to eat within a few score metres of the sett that Badgers don't need to waste energy searching more widely. Many householders feed urban Badgers, and we humans are renowned for our food waste. In Bristol, Badgers scavenge an amazing half of their total food intake. Given these advantages, it might seem strange that Badgers are not more abundant around human habitation. This is probably due to two major checks on population growth: urban traffic and limited sloping ground in which to build setts.

Below: Householders across Britain proactively feed Badgers from nearby setts, encouraging them into gardens and onto patios.

Eat seasonal, eat local

The Badger may be a member of the order Carnivora (meat eaters), but it is actually an unfussy omnivore, consuming both animals and plants. In another contradiction, Badgers have the digestive system of a carnivore, but the dentition of an omnivore. This is all intriguing enough, but what is even more fascinating about the Badger's diet is quite how widely it ranges. This is a mammal that flexes its taste buds to whatever the local environment provides.

As long ago as 1948, a grand old duke of Badger science called Ernest Neal described his study species as an 'opportunistic omnivore'. By this Neal meant that Badgers eat whatever is both available and edible, in a particular place on a particular night. Reams of research into Badger fodder confirms the overall principle that Badgers 'eat seasonal, eat local'.

Given the Badger's wide geographic distribution, it makes sense that an individual foraging in an Israeli desert in December dines differently from one wandering through a Swedish meadow in June. Yet even within a country, there can be remarkable dietary diversity. In Spain one study revealed badgers to be 77 per cent vegetarian; in another plants formed just 1 per cent of the mammal's intake.

Despite this variation, we can pick out the odd common thread. Typically, two-thirds of a Badger's consumption comprises animals, and one-third comes from plants in some form. Of the animal component, earthworms are the main prey, forming – across Western Europe as a whole – about one-quarter of overall intake. This average conceals considerable geographical variation, however. In southern Mediterranean countries, due to the arid conditions, earthworms stay far too deep in the soil to be extracted by Badgers. In Norway, Poland and northern England, however, mild and damp weather facilitates access to earthworms, which form around two-thirds of Badger diet.

Taken together, vertebrates (small mammals, amphibians, ground-nesting birds and more) and insects (wasps, bees, ants, beetles, moths, flies and so on) comprise another third of the diet. Cereals (particularly

Below: Bees may form a seasonally important component of the Badger diet, as determined from Badger poo.

Badgers consume a wide variety of vertebrates and invertebrates, including (**clockwise from top left**) Short-tailed Field Vole, Common Toad, wasps, and Earthworms.

maize) and fruit (notably many kinds of berry) comprise another quarter. The remaining proportion includes fish and crustaceans, slugs and snails, spiders and reptiles.

Young Rabbits are a staple food in southern Spain, while Badgers routinely scavenge Wild Boar carrion in eastern Europe's snowbound spring. It is thought that Hedgehogs are rare in western England because they are predated by the many Badgers that thrive there. To put all this in context, the Badger diet is much broader than that of fellow mustelids. Eighty per cent of an Otter's fare, for example, is fresh fish. In contrast, the Badger's principle appears to be 'if it moves, munch it'.

Below: Vegetables provide calories too: maize is a particular favourite.

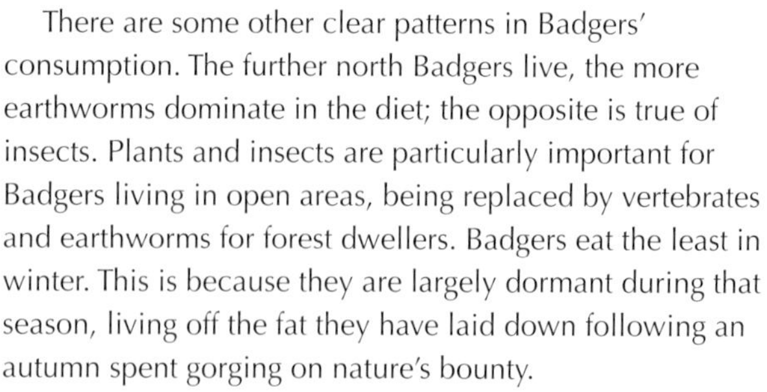

There are some other clear patterns in Badgers' consumption. The further north Badgers live, the more earthworms dominate in the diet; the opposite is true of insects. Plants and insects are particularly important for Badgers living in open areas, being replaced by vertebrates and earthworms for forest dwellers. Badgers eat the least in winter. This is because they are largely dormant during that season, living off the fat they have laid down following an autumn spent gorging on nature's bounty.

Given that Badgers eat whatever is seasonally available, it is no surprise that summer sees them catching the most insects, late summer involves a hike in cereal intake, and autumn features scrummaging for windfall apples and browsing for blackberries. Young mammals, particularly

Above: Other foodstuffs include blackberries (**left**), Hedgehogs (**top**) and young rabbits (**above**).

Rabbits, are snatched whenever their parents breed. Toads are typically eaten when they are easy to find. In France this is during March, when the amphibians gather in mating hordes at traditional ponds, and in September, when they travel to hibernation hollows.

Badgers also take advantage of occasional gluts. One Swiss study demonstrated that Badgers splashed out on Water Voles when they were abundant, but switched to earthworms when vole populations crashed. Badgers are particularly prone to binging on fruit when it is in season. They 'pig out' on plums and cherries in Switzerland, and olives in several Mediterranean countries, for example.

So what *don't* Badgers eat?

This flexibility is exactly the behaviour we should expect from a generalist feeder. But where do Badgers draw the line? What will they not eat? The short answer is... nothing.

The only potential foodstuff that Badgers seem to minimise, given its widespread availability, is fungi. But there is nothing that they unambiguously avoid. Badgers swallow leaves, grass and other green vegetation – although it is unclear whether these are accidentally consumed while foraging or deliberately munched to aid digestion. Badgers have even been recorded eating spicy curry and beer – reportedly with no unfortunate after-effects.

Finding food

Most of the Badger's fellow European mustelids mainly hunt by darting, lunging and pouncing, making the most of their lithe sinuosity and speed to catch vigorously mobile prey. Being foragers rather than hunters, Badgers could hardly be more different. There is method in their meandering. A Badger pads across pasture, its head swinging like a pendulum with its highly sensitive nose nuzzling within a couple of centimetres of the ground. Anything edible in its path will be rapidly sniffed out and, if stationary or nearly so, equally briskly snaffled up.

Worming around

Given how important they are to Badgers, let's examine earthworms in a bit more detail. You may think that any old worm will do, but that is not the case. Of Europe's many earthworm species (hands up if you knew there were 27 in Britain alone), Badgers are particularly partial to a single species, the Lobworm or Common Earthworm *Lumbricus terrestris*. This is one of Europe's largest earthworms, sometimes reaching 25cm (10in) in length. With its reddish-brown back and yellowish undersides, this is the worm that you are most likely to find in your garden, particularly after rain when it emerges to feast on decaying leaves.

There are three reasons for this love of Lobworms. Lobworms offer Badgers a good 'return' on the 'investment' of energy involved in capturing them. Earthworms are relatively easy to find and catch. Even better, each offers plentiful 'meat'; as a typical Lobworm weighs 10g (¼oz), a Badger can meet its entire daily energy needs by chomping 175 of them in a night. Finally and crucially, Lobworms are also relatively predictable creatures, emerging in abundance on nights that are warm, calm and damp.

In the UK and Republic of Ireland at least, this taste for worms largely explains Badgers' preference for living close to fields grazed by livestock. Pasture flourishes on rather moist soils ideal for worms. As much as 5 tonnes of worms may reside in the earth below every football pitch-sized field. Short grass nibbled by cattle or sheep enables Badgers to quickly detect worms that are above ground without the need for digging. This fast-food bar saves Badgers time and energy.

Above: Badgers love feeding in grazing pastures as the short, cow-nibbled grass makes it easy to detect worms straying above ground.

The keys to a Badger's foraging strategy are to cover an area thoroughly, and to pick up numerous bite-sized morsels, rather than a few big-ticket items. Once a Badger has arrived at a favoured feeding ground, it settles down to search, beetling around until it is satisfied that it has detected everything worth gobbling that particular night. Then, and only then, does it move on.

The distance travelled by a Badger each night varies with two factors: the size of its home range and the season. The bigger the range, the longer the journey a Badger must make to find sufficient food. This is not because Badgers need to travel great distances to be sure of getting enough food. Instead it makes sense to stock up during these seasons of plenty. An overall average, across Europe and the year as a whole, would be a nightly perambulation of around 4km (2½ miles).

Most food items are either lying on the ground or buried just below the surface. Prey above ground is gripped by the Badger's incisors and flipped up into its mouth. Subterranean

Below: Badgers forage by swinging their nose low over the ground, sniffing for prey and gobbling anything suitable that they find.

Above: Insects hiding in a tree stump are detected by scent then ripped out with sharp claws.

prey demands greater effort, typically requiring the Badger to root simultaneously with snout and long-clawed forepaws to extract the victim from concealment. Such digging leaves highly visible 'snuffle holes' of a few inches in diameter. Find a field spotted with these and you can be sure to have discovered prime Badger terrain.

Variations on this general approach reflect the particular characteristics of the prey. Badgers dig deep holes to reach young Rabbits sheltering in their warren. They rip open the bark of trees to get at a bee's nest. Toads (with their toxic skin) and Hedgehogs (those spines!) are flipped onto their backs, 'unzipped' and their insides chewed; the outer layer is left untouched. The stalks of cereals such as maize are knocked down and stood upon. This enables a Badger to leisurely nibble the calorie-filled heads. On occasion Badgers even climb trees to harvest fruit or slugs. Whatever needs doing to get at their supper, Badgers will do it.

Below: Even invertebrates living above ground are not free of attention from foraging Badgers.

Given that Badgers live together, you might think that they work together to find food. Once again, Badgers defy our expectations. Badgers not only forage alone, but also seem to deliberately avoid one another. Tim Roper tells of Badgers arriving at feeding spots within seconds of a fellow group member departing, as if they had been lurking shyly in the shadows waiting for the coast to become clear. The only circumstance when Badgers eat in a group is when there is enough food for all. In a field of ripened maize, for example, the attitude seems to be the more the merrier, even if 'more' includes Badgers from neighbouring groups. In such gluts there is nothing to be gained by fighting off other Badgers.

Above: There is some evidence that Badgers may predate domestic chickens.

Feeling thirsty?

Even scientists who have studied Badgers for years have rarely seen them drink. Indeed, they occur both in actual deserts (as in Israel) and in arid parts of temperate countries (such as England's chalk downland, where surface water is absent). This suggests that Badgers get all the liquid they need from their food, perhaps with the odd top-up from a temporary puddle. This strategy is not without its limitations, however. In unusually dry periods Badgers may not be able to digest the cereals that (in some parts of the world) form a sizeable proportion of their late-summer diet. In terms of rainfall, lean times mean lean Badgers.

Below: Badgers get almost all the water they need from their food, but occasionally top up by drinking from natural and artificial water sources.

Game, sett and match

All British mustelids use some form of concealed den, usually underground. Most purloin existing burrows made by other mammals, expanding them as necessary so they can fit inside. Only the Badger routinely excavates its home sweet home. And what a home it is! The Badger is up there with the Mole and Rabbit when it comes to the complexity of its subterranean domain. Indeed, the Eurasian badger's sett is without peer in terms of size and age – even among fellow badger species. Badgers, simply, are master builders.

The best way to tell whether Badgers occur near where you live is to go out looking for a sett. However, Badger burrows can look a bit like Rabbit warrens and Fox dens, so it pays to learn the tell-tale signs in advance. More on these clues later, but for now here are the headline trademarks of a sett.

First, the sett is always on a slope, and in soil that is easily dug and well drained (chalk is fine, clay is not). Then look for bare earth, usually under a tree or other vegetation cover, with well-worn paths radiating outwards. If you spy several large holes that are about the size of a football, but wider than they are high (that is, shaped like half an oval), then you should be onto something.

Well-established setts typically have a neat spoil heap of 'builder's rubble' outside, which comprises soil and, uniquely among Europe's burrowing mammals, stones. The biggest, oldest setts may have 80 entrances and 40–50 tonnes of spoil. You may also spot a tree trunk that has been scratched bare of bark, and clumps of dried grass which are the Badger's bedding. To be absolutely sure, get on your hands and knees like a scene-of-crime officer and look for the ultimate giveaways: Badger pawprints or hairs.

To track down Badgers, look for clues such as hairs caught on barbed wire (**above**), tracks showing the imprints of long claws (**below**) and spoil from recently dug holes (**below right**).

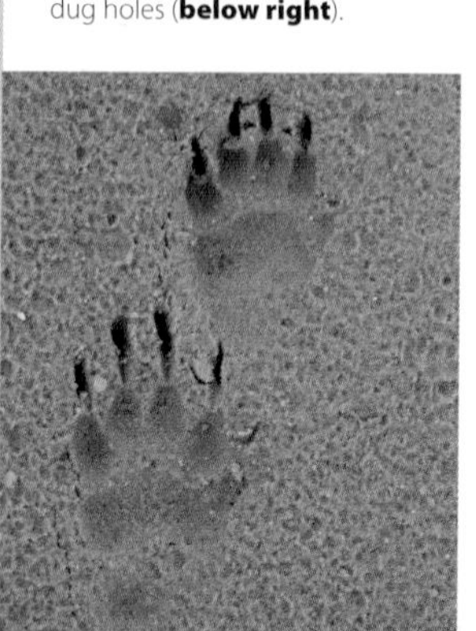

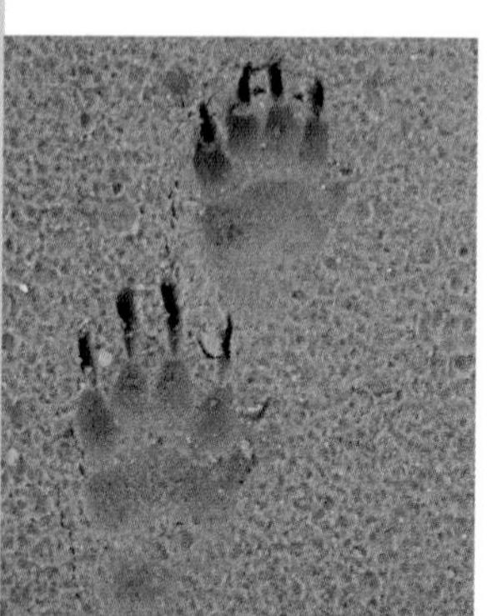

Above: Setts typically have a number of entrances spread across a large area of bare earth. Critically, setts are located on sloping ground.

Going underground

So much for the external signs of a sett. As with a good book, what really matters is what it is like on the inside – and like books, no two setts are ever the same. For sure, all have tunnels descending from the surface, enabling the owners to move between chambers where they settle down. All setts also contain one or more latrines, which are probably used only by nursing females or during winter, when Badgers are largely confined underground.

Apart from these common features, setts differ considerably in their vital statistics. They may be subterranean bungalows or have up to three storeys of tunnels and chambers. The number and length of tunnels varies hugely – from a single metre-long thoroughfare to 94 subways with a combined length of 310m (1,017ft). Chambers vary in number (up to 40) and size, but are usually only just big enough to fit one or occasionally two Badgers.

Over months, years, decades and even centuries – the oldest recorded sett had 475 years of continuous occupation – Badgers have clearly invested Herculean effort in construction. As natural-history writer Patrick Barkham expresses it in his delightful book *Badgerlands*, 'the badger is fêted for its fortresses'. Barkham goes on to explain the origin of the word sett. 'The collective noun for a group of badgers', Barkham writes, 'is a "cete", which probably comes from the Latin *coetus*, meaning "assembly" or "coming together"'.

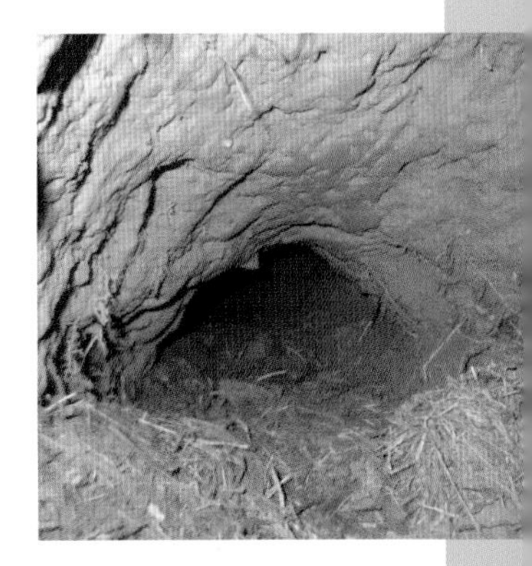

Badger holes are football size, but shaped like half an oval (**above**): a snug fit around a Badger (**below**).

Setts: why bother?

The sett merits the investment of construction because Badgers spend 70 per cent of their lives there. The sett is the focal point of a Badger group. It is both the safe haven where members hide from predators and shelter from the elements, and the shared space where they rest between foraging sorties, sleep, consolidate group bonds and breed.

Sleeping chambers typically contain up to 15kg (33lb) of insulating bedding, usually straw, dry grass or Bracken. In spring fresh greenery also features, typically Wild Garlic and Bluebells. Individuals towards the top of the sett's hierarchy assume responsibility for collecting its bedding, which they drag back to the sett from up to 100m (330ft) away. This can become a labour of love. Over the course of a single night one diligent Badger was watched gathering 30 bundles of bedding, weighing more than 1kg (2¼lb) in total. Badgers move sleeping chamber every couple of days and regularly air bedding, in both cases presumably to prevent the build-up of parasites.

Above and below: Badgers insulate the sleeping chambers of their sett with bedding such as straw and dry grass. Trails of dropped vegetation typically lead back to the sett.

Holiday homes

Pretty much every study of Badgers anywhere starts with the sett. Or, rather, *setts*. Most territories contain more than one. The lower the population density of Badgers, the larger each group's territory and the more setts it contains. A single group has been recorded using as many as 40 setts; in Norway the average is 13 setts per group and in Spain it is double that. A typical number in densely populated southern England, however, would be 3–6 setts.

Even for a group with just three setts, it is not surprising to learn that the refuges are neither used equally nor serve the same purpose. Main setts are permanently occupied, used for breeding and have several entrances. Scientists studying Badgers have identified three further types of sanctuary: 'outliers', 'annexes' and 'subsidiary setts'. Outliers usually have a single entrance, are sporadically occupied and, as their name suggests, lie towards the fringes of a territory. Annexes are basically spillovers from the main sett, have a few entrances and are regularly used. Subsidiary setts are like an annex, but are located in typical outlier terrain.

Sometimes Badgers shun setts, preferring to sleep above ground. This is particularly the case during summer in northern countries. In southern countries such as Portugal, outside slumber may be commonplace: members of one group were found to sleep above ground 40 per cent of the time.

Below left: Badgers often locate territories in Bluebell woodlands, using fresh flowers and leaves as bedding.

Below right: In addition to their main sett, Badger groups may have 'outlier', 'annex' or 'subsidiary' setts, which are used less frequently or by fewer animals.

Activity cycles

Above: Badgers typically emerge from the sett between sunset and two hours after nightfall (**left**). They do so tentatively, sniffing the air until they are sure danger is absent (**right**).

The fact that Badgers spend nearly three-quarters of their lives underground is the main reason why so few of us have seen the species other than as a roadside corpse. The other reason is that the fraction of a Badger's life that it spends in the open air coincides with night, when we humans largely treat ourselves to some shut-eye.

Badgers are almost exclusively nocturnal. For most of the year they emerge from the sett at dusk or later, timing their activity for when favoured prey such as Lobworms is out and about. In northern countries in particular, however, short summer nights mean that Badgers cannot rely on the hours of darkness alone to provide sufficient time to sate their hunger. This means that it is not uncommon to see Badgers emerge above ground well before the summer sun sets.

Below: An adult Badger leaving the sett at dusk.

A quivering nose cautiously sniffing the air is the first sign of a Badger's exit from its slumber pad. Once the Badger is happy that the coast is clear, it fully extracts itself and typically sets about grooming. Other Badgers trust the advance guard's judgement and exit the sett with less caution. Social activities such as mutual grooming and scent-marking follow, before group members trundle their separate ways and start foraging. During the night they may hunker down in a sett for a rest, or stay out until shortly before dawn's first glimmer.

Badgers' seasonal activity cycle is closely linked to climate. In Iberia mild winters enable Badgers to be consistently active year round. In Russia, where winters may be ferocious, Badgers spend as much as half the year wholly underground. In the UK they are particularly active from spring to autumn, and relatively inactive during winter. During the coldest months the entire group may remain in its sett for several days at a time. As and when they emerge, the Badgers may be active only for a few minutes. In Poland and Switzerland they are most active in summer and least active in winter.

While undoubtedly long, the periods of inactivity are better described as torpor (drowsy lethargy) than as hibernation. For sure, Badgers survive winter by living off fat reserves built up during the autumn glut rather than foraging anew. They also lower both their metabolic rate and body temperature to help them save energy – but not low enough for them to enter the deep sleep and prolonged immobility of true hibernation.

Above: The first animal to venture out of the sett is typically a dominant adult. Others quickly follow suit.

Below: Social activities are an important component in the early part of the night. These include mutual grooming and play.

Badgers
Moch daear

Threats and Conservation

If it were not for living alongside *Homo sapiens,* Badgers would generally be having a fine old time of it. Almost all of the threats they currently face are of human origin or are being made worse by humans.

Clearly, ebbs in food supply – notably during droughts – are a natural phenomenon that may kill off Badgers. However, even here it could be argued that these events are exacerbated by humans, given that climate change caused by our addiction to fossil fuels is thought to increase the frequency of arid summers. There is much more on our impact on Badgers later in this chapter. First, let's run through the unequivocally natural threats that they must counter.

Natural born killers

Across its range the Badger has a few natural predators. In continental Europe, Badger remains have been found in the stomachs or scat of the Grey Wolf and Eurasian Lynx. There are reports of Brown Bears, Wolverines and Eagle Owls attacking Badgers. Nevertheless, the evidence for Badger predation is neither strong nor widespread, and the overall conclusion is that Badgers have few mortal enemies.

Opposite: The biggest danger that Badgers face is living close to humans.

Below: Grey Wolf is one of the few natural predators of Badgers.

Other known or reported predators of Badgers include Eagle Owl (**above**) Eurasian Lynx (**top right**) and Brown Bear (**right**).

This is certainly true in the UK and Republic of Ireland. When Badgers returned to the British Isles at the end of the ice ages, about 10,000 years ago, natural predators may have included the quartet of carnivorous mammals mentioned above. However, all four have long been extinct there. Of meat eaters that remain, only the Red Fox, Golden Eagle and Common Buzzard are of sufficient size to down a Badger. Even then the victim would have to be either an adult that was sick or wounded, or a cub.

By and large, the mammal that a Badger actually most needs to fear is none other than fellow Badgers. Fighting is commonplace and such scraps can end in death. Even if the loser does not suffer fatal injuries, death may result indirectly: the Badger's wounds may reduce its ability to feed or lead to it blundering across a road into the path of an oncoming vehicle. Moreover, as we have learnt, dominant female Badgers may dispatch the cubs of subordinate sows to remove the competition to their own cubs.

Right: Common Buzzards are theoretically big enough to down an injured or young Badger.

Do Badgers and Foxes compete?

Badgers and Red Foxes are similar-sized mammals that both breed underground and eat a variety of food. Theoretically, there is potential for competition between the two species in both subterranean space and diet. In reality, however, competition seems all but absent, in Britain at least. The two species tolerate each other, even to the extent of their respective females raising young in the same underground retreat.

In normal conditions rural Badgers and Foxes are unlikely to vie for the same prey. Badgers prefer Lobworms, Foxes small mammals. On suitably damp, mild nights, there should be more than enough worms for both. When Badgers eat mice, Rabbits or voles, they tend to target youngsters at the nest, whereas Foxes are adept at hunting adults that venture away from their hiding places. In abnormal weather, such as drought, food may be scarce for both predators, so competition is more likely to occur. In urban areas both species scavenge a sizeable chunk of their food. However, the source of that food, us humans, is such a prolific waster of edibles that there is more than enough to go round.

Diseases and parasites

Above: The Badger's preferred foraging method – nose close to the ground and rooting in the soil – exposes it to pathogens and parasites.

Snuffling along with their noses to the ground and rooting around in soil readily exposes Badgers to ingesting parasites and catching diseases. Sleeping communally in the confines of a sett – with high humidity and moderate temperatures – enables both types of affliction to spread rapidly. By trotting into a neighbouring territory, an infected Badger can unwittingly spread a malady.

Three sets of parasites make their homes in Badger fur: fleas, lice and ticks. Research by biologist Tim Roper indicates that most Badgers suffer from all of these, and all Badgers endure at least one. Healthy Badgers have few parasites. This suggests that lengthy grooming sessions are worthwhile. Given that grooming is usually a Badger's first activity when it emerges from a sett, however, it suggests that Badger bedding may be heaving with invertebrates. Weakened, diseased or starving Badgers have a higher number of creepy crawlies living on them than do healthy Badgers. This is presumably because the unfortunate individual is too debilitated to groom itself properly and too

Below: Removing unwanted and potentially dangerous parasites – fleas, lice, ticks and the like – is an important daily activity.

low in the group's hierarchy for another animal to help out.

A score of infections – whether bacterial, viral or fungal – have been recorded in Badgers. In Europe, with the exception of the UK and Republic of Ireland (where the disease is absent), rabies is by far the most significant cause of death. Where it enters a local population, rabies can have a devastating impact, killing up to 90 per cent of animals. It spreads rapidly through Badgers biting or licking an infected animal. Fortunately, the virus is not common in Badgers: no more than 2 per cent of rabies cases in continental Europe's wild animals occur in Badgers, compared with 90 per cent in Foxes.

Bovine tuberculosis kills Badgers too

For cattle farmers in the UK and Republic of Ireland, Badgers are not very popular when it comes to bovine tuberculosis (bTB). This is because they are a 'reservoir' for a disease that can result in significant numbers of cows being prematurely slaughtered. More of this later; for now, let's consider the physical impact that bTB has on individual Badgers.

The way in which bTB manifests itself in Badgers is pretty complex, and scientists have yet to work out why certain differences occur. Some Badgers are exposed to bTB, but somehow avoid infection. Others are infected, but are not infectious so cannot transmit the disease further. Some are contagious yet show no symptoms. Yet others have the full monty: they are weak, display a full set of symptoms and are rampantly infectious.

Where TB attacks a Badger's internal organs, it creates pale lesions filled with bacteria. The disease is mostly transmitted between Badgers by coughing or sneezing. Other methods of transfer depend on where the lesions are located, but include licking, biting or getting too close to infected wee or poo, particularly at latrines.

Given this muddle, working out how significant a threat bTB is to Badgers is far from easy. The long-term study at Woodchester Park (Gloucestershire) suggests that 80 per cent of Badgers never contract bTB, with just 5 per cent becoming infectious and 2–3 per cent super-infectious. The same study found that Badgers that reach the most advanced stage of bTB are twice as likely to die as healthy Badgers.

Below: This Badger is being released after being vaccinated against bovine TB by the Cheshire Wildlife Trust.

On the road

Many of us have only seen Badgers as they sprawl lifeless on a roadside verge, having been obliterated by a vehicle. Indeed, most of us have probably seen numerous Badgers this way. This drops a heavy hint about the scale of the impact that road traffic has on Badger populations. For sure, vehicles are the biggest cause of Badger deaths. Across Britain, for example, at least 20 per cent of adult Badgers die this way *each year*: this is half of all adult Badger deaths. Fortunately, the country's overall Badger population is probably sufficiently high to cope with such losses without entering decline. In Denmark, however, as many Badgers are killed each year on roads as cubs are born. The result is a shrinking population. The same trend, with the same cause, is predicted for Sweden by 2050.

Did you spot that the statistics above relate to *adult* Badgers? Deaths on the roads almost exclusively relate to adults – and peak in spring. Given that this is the main mating season, it figures that adults are killed while moving between territories in search of breeding opportunities. What starts out as a race for life ends in death.

Above: One-fifth of all adult Badger deaths in Britain are due to motor vehicles.

Road signs in areas of high Badger density encourage motorists to slow down (**above**), but such warnings are not always heeded (**right**).

Habitat change

Above: New roads cause Badgers two problems. They potentially fragment long-established territories and they bring vehicles into contact with new groups of Badgers.

The biggest threat to animals worldwide is human change to habitats. Deforestation is the most publicised example, but agricultural intensification, drainage of waterways, road building, urbanisation and overgrazing feature in a mixed bag of degradation and destruction. Remarkably, Badgers seem resilient to pretty much all of this.

We know already that urban Badgers do very well indeed. Due to the abundance of human food waste, territories can be small and population densities high. Because of Badgers' all-embracing diet in rural areas, they are not deterred by most agricultural landscapes. Even if territories have to be large, their Badgers can always find something to eat. The main constraint on farmland is the infrequency of slopes into which to excavate a sett. Badgers are only usually deterred by relentlessly flat arable plains.

Roads themselves, as opposed to the traffic that speeds along them, tear habitats suitable for Badgers into fragments. In the case of dual carriageways and bigger roads, they also present tarmac barriers to Badger movement. The first problem constrains territory size. The second makes movement between territories or one-way dispersal a risky business. Nevertheless Badgers somehow cope with both sets of problems. If it weren't for the vehicles coursing their length, Badgers and roads would get on fine.

Above: In some areas, Badgers are still 'baited', being set upon by trained dogs.

Hunting

Road traffic and habitat change are inadvertent human threats to Badgers. Now it is time to peer into the morally murkier world of wholly intentional pressures, those where we humans actively wish Badgers harm. Across much of Europe, hunting has long been a major threat.

In Germany alone, some 50,000 Badgers are legally shot each year, many of them in areas where they damage crops. In Switzerland Badger meat is considered a delicacy, so hunting is common. In Romania hunting is illegal yet still occurs, principally to extract the rich fat that Badgers build up during an autumn of profitable foraging. In contrast, across the UK and Republic of Ireland, Badgers are not and have not been routinely hunted for food.

Relentless persecution

Across centuries, and from Britain to Bulgaria, we humans have persecuted Badgers. We have dug them out of their setts, 'baited' them by setting specially bred dogs on them, killed them with snares, poisoned them and shot them. There is surely no other mammal that we have so intensively hounded. We examine in more detail this unsavoury dimension to the relationship between people and Badgers in the following chapter. Here we focus on the demographic impact of our relentless badgering on the mammal's populations.

Below: Badgers are also still trapped in snares. If they survive, the wounds can run deep.

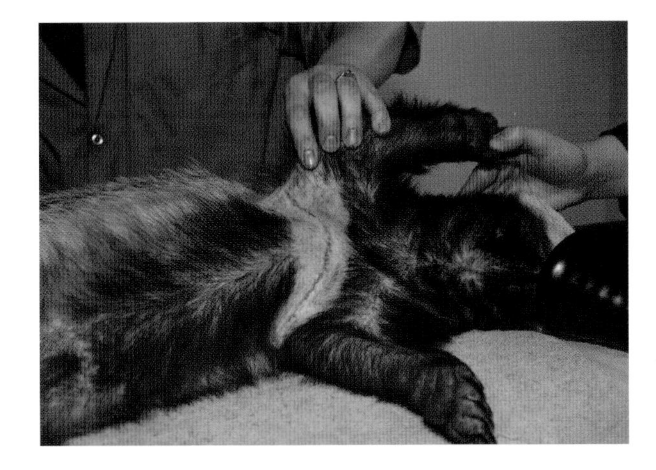

Left: This Badger, at Taunton Wildlife Hospital, is lucky. It sustained chest wounds from a snare, but has been sewn up under anaesthetic.

In the UK, persecuting Badgers for 'sport' or 'vermin control' is illegal. This makes it tricky to work out the impact of such activities on Badger numbers.

As for digging and baiting, two sets of research – one national, one local – revealed very different patterns. A major survey during the 1980s found that one-tenth of all setts were 'dug' every year. The following decade a survey in Yorkshire made the even more concerning suggestion that setts were dug twice *per year* on average. It also revealed that the more remote the sett, the greater the chance of it being attacked – presumably because the diggers felt they were less likely to be caught in the act.

Below: The culprits have tried hard to cover their tracks, but it is nevertheless clear that this Badger sett has been dug over.

Stopping Badgers spreading disease

Earlier in this chapter we learnt about how diseases such as rabies and bTB can kill infected Badgers. While the infections might affect local populations, they rarely, if ever, have wider influence. In contrast, part of our attempts to eradicate those diseases from the landscape is intended to have a profound impact on Badger populations and distribution in target areas.

Rabies was a real problem in continental Europe in the 1960s and '70s. To remove the disease from wild animals, Fox earths and Badger setts were routinely and widely gassed. In Switzerland the Badger population may have shrunk by as much as 80 per cent within 20 years. By the 1980s, however, the focus of rabies control shifted from killing the mammalian reservoirs of the infection to vaccinating them against the disease. This proved vigorously successful. Many continental European countries are now free from rabies. In Switzerland and probably more widely, Badger numbers are now back to their pre-cull levels.

Below: Interaction between Badgers and livestock, such as this German Badger chasing a sheep, has the potential to spread disease.

Conservation

Left: Demonstrators wearing Badger masks participate in a 'flashmob' in London in October 2012. They were protesting against the UK Government's pilot cull of Badgers.

Looking at these threats in the round, traffic is the most significant limitation on Badger populations, particularly in western and northern Europe. Moreover, it is only likely to increase in the future: expect to see many more prostrate grey bundles littering the roadside. Hunting has a considerable impact at a national level in countries where it is permitted, particularly in central and eastern Europe. Disease is not a big issue; rabies has been vanquished from Badgers, while bTB only affects populations locally. Culling and illegal persecution can be locally important, where they occur.

The Badger is distributed over a very wide area of Europe and Asia, with its total European population being at least 1.5 million. Overall, Badger numbers seem to be stable or increasing – although their nocturnal habits and social structure do not make surveying them easy. Populations overall are apparently able to withstand the current level of losses from all the threats discussed above. Badger populations have bounced back from a myriad of pressures. This means that the Badger's situation is not of overall conservation concern.

The International Union for the Conservation of Nature (IUCN), which routinely assesses the global conservation status of many groups of animals, rates the Badger as being of Least Concern. International trade in Badgers or their parts is not considered to threaten the species sufficiently

Below: Many people are fond of Badgers. This two-week old cub narrowly survived floods, and was being hand-fed by workers at Secret World Wildlife Rescue in Somerset, UK.

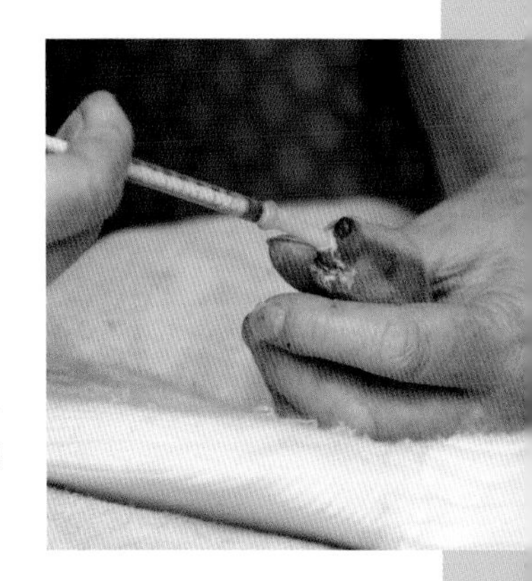

Above: A Badger-cull protestor outside the Department for Environment, Food and Rural Affairs in London, in May 2013.

Below: Temporarily catching Badgers for scientific study is legitimate. This cub is undergoing a medical examination at the long-running study site of Wytham Woods, Oxfordshire, UK.

for it to be part of the international agreement that covers such activities (the Convention on International Trade in Endangered Species of Wild Fauna and Flora, or the more accessible CITES for short).

None of this means that the Badger is off the radar of conservationists. Many of the countries in which Badgers occur have signed the Convention on the Conservation of European Wildlife and Natural Habitats (in shorthand, the Bern Convention). This binds governments into a deal. They may allow hunting/culling of Badgers within their country – but only if they regulate its level to 'keep its populations out of danger'. So Badgers may be hunted, but under careful oversight and according to conditions set nationally.

Moreover, in some countries concern about the Badger's status has led to proactive conservation measures being taken. In the Netherlands action has varied from providing underpasses below roads to reduce mortality from traffic, to selective reintroductions to formerly occupied areas. Badgers have also been reintroduced to a site near Milan, Italy.

Brock and the beak

The UK and Netherlands have gone further than any other country, their respective statute book each containing legislation designed to protect Badgers. In the Netherlands Badgers and their setts enjoy protection through the 1998 law on the Protection of Flora and Fauna. Effective in England, Scotland and Wales, the Protection of Badgers Act 1992 is the most recent of several laws with this aim. The Act makes it illegal to interfere with a sett. 'Interference' ranges from deliberately disturbing its meline inhabitants to destroying their home, and from selling a live Badger to possessing a dead one unless legally 'obtained'. It is also against the law to build within a certain distance of a sett, or even to lop down adjacent trees. Badgers and their setts are protected in Northern Ireland under the Wildlife (Northern Ireland) Order 1985.

From a UK Badger's perspective this sounds good. There are a couple of downsides, however. First, any regulation is only as effective as its enforcement – which means catching lawbreakers. Crimes against Badgers, and other wildlife, can be difficult to enforce because they often occur on private land in remote locations away from public scrutiny. Second, the Protection of Badgers Act permits action to be taken against problem setts, provided a government licence is granted. A 'problem sett' might be one whose occupants carry disease or raid crops, or which is located on the site of proposed construction of some kind, from a road to a building.

Licences are not granted willy-nilly. For example, Natural England – the public body in charge of such things in England – expects that most developers should be able to avoid harming Badgers by adjusting their plans. For road builders, for example, this might include flanking a new carriageway with Badger-proof fencing, or constructing underpasses so that Badgers may cross beneath roads rather than risk the vehicular gauntlet above.

Applying for a licence is expected to be a 'last resort'. Moreover, Natural England will only consider licensing the activity if it would be impossible to avoid disturbing the animals and if there is a robust compensation plan in place. 'Compensation' might mean providing an alternative sett within the Badger group's territory, or improving surrounding habitat. Conditions may also be set, such as timing the work to avoid the months when sows are pregnant or nursing cubs.

Natural England currently considers around 20 applications for licences per week, and grants perhaps 15 of these. Requests now come at roughly double the rate of the 1990s – particularly in relation to urban Badgers. This is the result of the mustelid's increasing population and spreading distribution, both of which increase the likelihood of pitting Badger interests against ours.

Below: Road-builders are increasingly required to avoid harming Badgers by providing underpasses (**left**) or badger-proof fencing (**right**).

Badgers and people

Across the UK and Republic of Ireland at least, Badgers are arguably more engrained than any other mammal. No other furry creature is such a prominent architect, shaping the countryside through its Gorgonzola setts. No other four-legged animal is immortalised so abundantly in the names of human habitations such as Brocklehurst and Eshott Brocks. Badgers are embedded in our literary culture as deeply as any other mammal, *Fantastic Mr Fox* and *Tarka the Otter* included. Yet our attitudes to Badgers are complex to the point of chaos.

Some people live with Badgers; others loathe them. Some people feed Badgers; others bait them. Some people cherish Badgers as fictional characters; others chastise them as cattle infectors. Some people spend evenings watching live Badgers; others pass nights pursuing their demise. Some people identify with Badgers, lauding their tenacity; others harass them, wanting them gone from their country. Some people use Badger images to sell things; others sell actual Badger parts. Can there be any other mammal that evokes such contrasting perspectives? Is the Badger the Marmite of mammals?

Opposite: Are our attitudes to Badgers now giving them a fair wind... or a foul one?

Below: We are able to live amicably side-by-side with Badgers, which are welcomed into many gardens across the UK.

Badgers in language and landscape

Above: Badgers have featured on stamps from countries including the UK, Poland and the former Soviet Union.

Several mammals have wriggled or slunk their way into the English language. Criminals and schoolchildren alike sometimes 'rat' on one another. Jimi Hendrix sang about a 'Foxy Lady', and we may be 'foxed' by a dilemma. We 'ferret' around, looking for something we have lost. And we 'badger' someone to get them to do what we want. The derivation of this verb reflects one saddening dimension of our relationship with Badgers. Throughout history we have badgered – harassed, persecuted – Brock by forcing them out of their homes, into no-win fights with dogs and worse.

Badgers' impact on the landscape of Britain, at least, has been no less significant through the evolution of place names than through the architecture of its setts. At least 140 British locations nod towards the local existence of Badgers, from the now resolutely urban Brockley in south-east London to defiantly rural Brockenhurst in the New Forest, from Hertfordshire's Broxbourne to Badgers Mount in Kent.

Also in Britain, Badgers have morphed into traditional surnames. In Suffolk alone there are coats of arms depicting Badgers for Brokes and Brooks. On a personal level I have admired the artwork of a Brockie, worked with a Brocklehurst and lived adjacent to the parliamentary constituency of a Brokenshire, while my wife schooled with a Brocklebank. Such names even make it into our literary heritage. In *Jane Eyre*, Charlotte Brontë's 19th-century classic, Mr Brocklehurst is the hypocritical supervisor of a boarding school for orphaned girls.

Right: From Broxbourne to Mr Brocklehurst, Britons have enshrined Badgers in our place names, surnames and literature.

Using Badgers

Being instantly recognisable, suitable for black-and-white reproduction, and associated with esteemed human characteristics of solidity and toughness, it is not too surprising that Badgers have served well as marketing tools across many branches of commerce.

In Britain, for example, Brits sup ale from Dorset's Badger Brewery. Little Badger makes knitwear for children. Badger Learning is an educational publisher and retailer. Badger Associates are recruitment consultants in East Anglia. A major supermarket chain – Tesco's – has Badgers on its coat of arms above the motto *Mercatores coenascent.* (The words are intended to mean 'Traders rise up together', but are also a play on the surname of the chain's founder, Sir Jack Cohen.)

It is not merely in Britain that Badgers sell. The Badger Company is a Netherlands-based engineering firm. In the United States customers of Badger Balm buy organic skin-care products; the company's logo is an American Badger. Badger's Brook estate in Australia's Yarra Valley produces eminently quaffable wines. The German family Thurn and Taxis, which started the European postal system, has a Badger at the core of its coat of arms. The French cycling champion Bernard Hinault, who won the Tour de France five times, was nicknamed *le Blaireau* ('the Badger') on grounds of his courage and combative nature.

Above: One brewery in Dorset named itself after Badgers.

Below: Somewhat less savoury, some shaving brushes are made from Badger hair.

The Badger's qualities serve non-profit uses as well. Most famously in Britain, the Badger serves as logo of The Wildlife Trusts, conservation charities that protect nature across the country. Road signs pointing you towards local reserves – a Badger face on a brown background – are a familiar sight. Skybadger helps families of children with disabilities, and won the *Guardian* newspaper's Charity Award in 2014. *The Badger* is the University of Sussex Student Union's free newspaper.

A different take on using Badgers can be seen in the aggregation of Badger guard hairs to form shaving brushes. Connoisseurs advise that the plus points of Badger hair are its water-absorption qualities (enabling a wet shave), fine tips (key to a creamy lather) and softness (for smooth defoliation). A typical brush comprises 14,000 hairs. Such is the reverence with which these shaving brushes are held that there are actually three classes of brush made from different parts of the coat. In order of increasing quality, these come from the Badger's belly, back and neck.

Producers say that bristles are taken from captive-bred animals, typically reared in China – a claim disputed by the Badger Trust, a British wildlife charity, which has provided evidence that they come from wild-killed individuals. Having previously been the preserve of the wealthy on account of their significant cost, Badger brushes are now widely available. Indeed, a well-known Internet retailer has a somewhat disturbingly wide selection.

Above: Badgers make for a rather disconcerting (and distasteful) fashion accessory. This sporran is made from an American Badger.

Below: Badgers feature as food in several European countries. This is the cooked head of a roadkill Badger from Cornwall, UK.

Shaving brushes are far from the only commodity that Badgers have provided across centuries of human 'use' (many might prefer the term 'exploitation'). Some posh paintbrushes are made from Badger hair, which one retailer claims are 'ideal for varnish, topside and high-gloss marine coatings'. In 2012 designers Hermès and Jean-Paul Gaultier both sent models striding down the catwalk clad in Badger fur: the popstar Lady Gaga was criticised for wearing the former's Badger coat. Badger coats and, disconcertingly, heads have long been used as sporrans, the pouch draped over the front of a Scotsman's kilt.

There is evidence that Roman soldiers ate Badger and that Bronze Age settlers on the Outer Hebrides may have 'imported' Badgers from the mainland. Some uses persist today in various parts of the Badgers' range. In Romania hunting for Badger fat remains regular, as does shooting for meat in Alpine Switzerland, where Badger is considered a delicacy. An Austrian recipe involves braising Badger with laurel, peppercorns, salt, red wine and sour cream. Meanwhile, Badger becomes goulash in the Balkans or shish kebabs in Russia.

In Britain there is a tradition of making hams from Badger hocks. Some restaurant menus featured Badger dishes until the species became legally protected in the 1970s. Intrigued by this, natural-history writer Patrick Barkham dined on roadkill *Meles* as part of research for his book *Badgerlands*. Barkham described stir-fried Badger as 'very dense, chewy and unpleasantly strong… finishing my plate was proving to be a slow process'.

Badgers as friends

Very few Britons may have eaten Badger, but many have helped Badgers to eat. Throughout rural and urban Britain, householders put out food for 'their' Badgers. These diet supplements might include anything from cheese to chicken and grain to grapes. In a Brighton suburb where Badgers abound, more than a quarter of residents admit to feeding Badgers, with one in six doing so every night. In Bristol half of all food eaten by Badgers is scavenged, and a fair proportion of this will be intentionally provided. One local resident even featured on the BBC TV series *Autumnwatch* as she nourished Badgers that lived in the cemetery adjacent to her flat.

It is not just in Britain where Badgers are fed by well-meaning people. Badgers in Japan given a nightly feed of half a kilogram of carbohydrates such as noodles put on weight far more rapidly than those that forage entirely naturally.

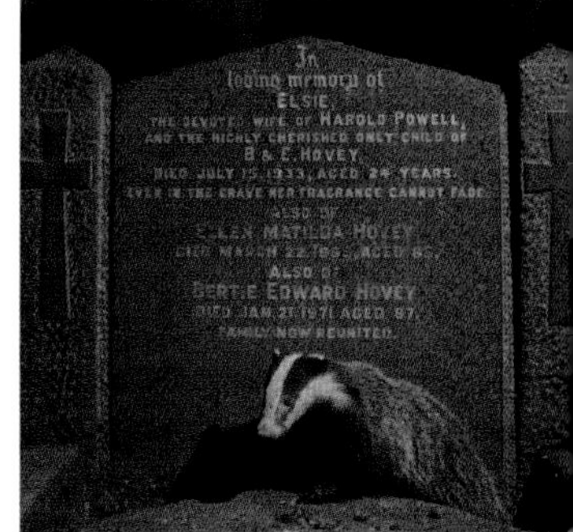

Above: Badgers in Bristol (UK) have featured on television wildlife programmes.

Left: Where they are welcomed into gardens, Badgers may become comparatively tame.

Above: Research on urban Badgers has revealed that significant food availability in towns enables them to have very small territories.

Little wonder that Badgers stick around: why waste energy shuffling and snuffling for miles each night when you can subsist on handouts within metres of home? Moreover, from the provisioner's perspective the attractions are obvious: a much-loved animal in your garden, every night. Artificial provision, however, is not without unintended consequences. It takes worryingly little for a Badger to be pushed over the narrow line from friend to foe.

Right: When Badgers are confident that they are not under threat, they can become very confiding.

Badgers as foes

The more habituated Badgers become to people and the more we provide edible reasons for them to live around our homes, the greater the prospect of conflict. Badgers may dig up lawns, which soon become unusable for the owners. Worse, Badgers may excavate setts beneath buildings that may become unstable and uninsurable.

Even if we do not proactively nourish Badgers, our agricultural and horticultural industries inadvertently provide them with potential dinner. Badgers regularly damage or destroy commercial produce ranging from cereals to grapes, and chickens to strawberries.

Farmers in England estimate that Badgers damage roughly 5 per cent of oat and maize crops, and lesser percentages of wheat and barley. Between one-fifth and two-thirds of English vineyards are troubled by Badgers, with average estimates of losses around the £500 mark. The financial impact of livestock predation by predators is thought to be £1–3 million per year. Although not economically significant for the sectors as a whole, many individual farmers and growers can ill afford such losses. We should not be surprised that ill feeling and persecution can result.

Overall, humans have a long and unflattering history of badgering Brock. Britons have done so since the 16th century. The rise of the country estate in the 18th century came hand in hand with an increase in the

Below: The juxtaposition of Badgers and humans can cause conflict, including damage to pasture (**left**) and commercial crops (**right**).

Above: Badger digging can also cause infrastructural damage, as could soon be the case beneath this estate road in Dumfries and Galloway (Scotland).

number of gamekeepers nationwide. For men entrusted with maximising their laird's bag, carnivores became incarnations of the devil – and Badgers became tainted with the same brush.

Across Britain some gamekeepers and landowners have treated Badgers as vermin, arguing that they predated game birds or lambs. It is true that Badgers occasionally kill chickens and, conceivably, could capture the youngsters of other game birds, too. The emphasis is on 'occasional', but this infrequency was not always recognised. Gamekeepers' methods of removing these mustelid 'pests' ranged from those that cause instantaneous death (shooting), to others that do not (snares, poisoning).

Right: Gamekeepers in some estates still set snares to catch Badgers.

Badgers as bait

In parts of Britain, there persists an indefensible practice that sends shockwaves through any community that loves or admires Badgers. In today's world, it is barely credible that there exist criminals who consider it 'good sport' to terrify, torture and ultimately kill Badgers. 'Baiting' Badgers involves forcing an individual out of its sett, whether by 'digging' or sending trained terriers after it. The Badger is then confined to an enclosed space and one or more purpose-bred fighting dogs are let loose. The Badger may put up a good fight –

deploying powerful limbs and jaws to good effect – but, ultimately, there can only be one outcome.

Unlike hunting, which is widespread throughout the Badger's range, baiting appears to be peculiar to parts of the United Kingdom and Republic of Ireland. It also has a disturbingly long history. Baiting can be traced back to medieval Britain, when rowdy taverns saw scenes of drunkenness, gambling and barbarity. In backyards, beer-fuelled bets were placed on how long a Badger would last or whether a hound would remain unscathed.

The nation's collective conscience only started to prick towards the end of the 18th century. In 1800 the first attempt was made to make baiting of Badgers (plus bears and bulls) illegal. It bit the dust, being shouted down by the press and Parliament. Early animal-rights activists such as Richard Martin MP were not deterred, however, and by 1835 successfully obtained a ban on Badger-baiting through the Cruelty to Wild Animals Act.

With laws, however, come loopholes. While the Act made it illegal to torture *wild* Badgers with dogs (ironically, to protect the canines from harm!), there was no such prohibition on *captive* victims. So all a pub landlord had to do to attract an excited rabble to his backyard was to procure a 'tame' one. It also remained legal to dig a Badger out of its sett, provided that dogs were not then set upon it.

Dog breeders even produced a short-legged, long-bodied dog design – the Dachshund – for the purpose of smelling, chasing and flushing out subterranean mammals. The primary quarry is captured in the dog's name: in German *Dachs* means Badger. Dachshunds, or 'sausage dogs', are certainly fit for purpose. In 2008 and against opposition in the form of breeds such as Pit Bulls, Rottweilers and Rhodesian Ridgebacks, the *Telegraph* even proclaimed the Dachshund to be the most aggressive canine breed.

Although Britain's Cruelty to Wild Animals Act may have reduced the incidence of Badger baiting, it did not eradicate the activity. There is plenty of evidence that it continues today, nearly 200 years and much Badger-protection legislation later. So who baited Badgers in Britain – and who still does so?

Below: Sausage dogs or Dachshunds have been bred specifically to chase animals such as Badgers through underground tunnels. *Dachs* is German for Badger.

Who are the baiters?

In *Badgerlands*, Patrick Barkham argues that the British Badger's downfall is that, unlike Red Deer or Red Fox, it was never hunted by the landowning aristocracy. For sure, Britain's gentry indulged in Badger *digging* during the 20th century, but captured animals were typically released. Granted, landowners expected their gamekeepers to *shoot* Badgers as vermin. However, unlike deer and Foxes (both considered noble creatures and worthy opponents), Britain's upper classes neither *hunted* Brock with horses and hounds, nor *pursued* it until death.

The ruling class's relative disinterest in Badgers left Britain's masses free to get their bloodsport kicks from baiting that particular mammal. Thus, says Barkham, Badgers became fair game for the working-class man. They remain so today, even if the majority of persecution occurs well under the radar. Anecdotal evidence suggests that northern England remains the stronghold of baiting, particularly on the outskirts of old industrial towns.

Unlike Fox-hunting, whose red-coated proponents are visibly and vocally proud of their pastime, Badger-baiting occurs far from mainstream society. Talking to people who identify a relationship between baiters and petty criminals from urban housing estates, Barkham argues that working-class families hand down practices and techniques from one generation to the next. He concludes that Badger baiting is a symptom of Britain's class-based system:

> *"Men who have very little power in their working lives seize control in another sphere. It may be unspeakably cruel, it may be an expression of our basest instincts – man as a bully, coward and thrill-seeker – but it is an expression of autonomy and freedom, and of one class's contempt for the laws made by another."*

Badgering baiters

Those in Britain who care about Badgers – and that is surely the majority of readers of this book – do not take the species' disturbance or deaths lying down. So who badgers baiters, diggers and gamekeepers engaged in illegal activities? The answer is an impressively broad and collaborative spectrum of bodies and individuals.

Sufficient illicit activity exists for Badger persecution to routinely feature in the top six priorities of the Partnership for Action against Wildlife Crime (PAW), a multi-interest group led by the Department for Environment Food & Rural Affairs (Defra). PAW oversees Operation Meles, an anti-Badger-crime partnership between respective UK police forces and key charities. The latter fall into two broad categories: broad-spectrum animal welfare or conservation groups such as Care for the Wild International and the League Against Cruel Sports; and Badger-specific charities, notably the Badger Trust, Scottish Badgers and the Badger Protection League.

PAW partners collaborate by gathering intelligence and targeting offenders. Much information comes from members of the public who spot evidence of wrongdoing while wandering the countryside. Headlines in the very first Operation Meles newsletter, in 2013, speak volumes about the ongoing frequency and breadth of Badger harassment: 'dog found in sett', 'dug sett discovered', 'suspended prison sentence', 'Badgers found in cages', and more.

Since 2010 several charities involved in Operation Meles, and more besides (including the RSPB and Wildlife Trusts), have battled against a very different type of Badger killing in England. This is the UK Government's cull of Badgers in parts of south-west England – the most controversial weapon in the struggle against bTB.

Below: Cattle pastures typically harbour abundant and accessible earthworms; little wonder that Badgers love foraging in such places.

Badgers and bTB

In 1971 a post-mortem on a dead Badger in England revealed that it had died of bovine tuberculosis (bTB). This provided the first evidence that this deadly bacterial infection – which the UK had been battling for decades – could make the leap from cattle to Badgers. It was also the first suggestion that wildlife could form a font of the disease, capable of reinfecting cattle. A shock wave reverberated through Britain's farming, scientific and conservation communities – and has not stopped since.

Cattle in crisis

Unlike its European counterparts, parts of the British cattle industry have endured a bTB problem for many decades. In the 1930s, 40 per cent of British dairy cows were infected. In 1950 a 'test-and-slaughter' regime was introduced. Animals that tested positive for bTB were immediately killed to prevent the infection from spreading further. By 1980, bTB infection had been reduced to less than one herd per thousand. By the 'noughties', however, bTB was once again widespread.

There are considerable differences in bTB rates across Britain. Scotland is officially TB-free. By 2010 there was nine times more bTB in England than in 1997, and more bTB in England alone than in the rest of Europe combined. In the ten years to 2014, 314,000 otherwise

Below: In 2010, there was more bTB in English cattle than in the rest of Europe combined.

healthy English cattle were prematurely slaughtered due to contracting bTB (with many more killed after contracting other diseases), and, according to Defra, the taxpayer has footed a bill of £500 million. In Wales there is considerable regional variation, with rates lowest in the north-west and north-east.

What have Badgers got to do with bTB?

There is a complex relationship between cattle and Badger infection, with the disease known to spread from cattle to cattle, from cattle to Badger and from Badger to cattle. Researchers believe that there must be reservoirs of bTB *outside* the cattle population that, in some areas, are reinfecting previously healthy herds.

Experience from other countries – bTB-infected Bison in Canada and Brushtail Possums in New Zealand – pointed the finger at wildlife. Unfortunately for Brock, and for those who love him, the evidence suggests that Badgers are indeed a significant potential source of infection. *How* Badgers and cattle infect one another remains a mystery that scientists are yet to unravel. Badgers may have got the disease in the first place by unwittingly consuming or inhaling infected cattle poo, wee or saliva. Badgers may return the 'compliment' through the same means.

Whatever the routes of transmission, the conclusion is clear. Eliminating bTB from British cattle involves removing it not just from herds, but from reservoirs of the disease in other species, too. It follows that any strategy seeking to eradicate bTB in domestic cattle also needs to remove it from wild Badgers. The question is: how?

Tackling bTB in Badgers

There are three broad approaches to tackling bTB: vaccinate animals so that they are immune to the infection; kill infected animals and wildlife 'reservoirs' (such as Badgers); or prevent cattle from coming into contact with those wildlife reservoirs. Each course of action has its pros and cons, its advocates and adversaries.

Badger culling may rarely have been out of the headlines since 2010, but its history stretches back to 1973, shortly after that first bTB-infected individual was discovered. For example, Badger setts were gassed over a wide area of south-west England from 1975 to 1982, until public opposition forced suspension of this tactic. A succession of independent reviews and advisory committees followed, each broadly charged with strengthening the understanding of the government of the day.

The most significant study in England was the Randomised Badger Control Trial, which ran from 1998 to 2005. The Independent Study Group on Cattle TB, which oversaw the trial, reported in 2007. The Group drew two key conclusions:

> *"First, while badgers are clearly a source of cattle TB, careful evaluation of our own and others' data indicates that badger culling can make no meaningful contribution to cattle TB control in Britain. Indeed, some policies under consideration are likely to make matters worse rather than better. Second... scientific findings indicate that the rising incidence of disease can be reversed, and geographical spread contained, by the rigid application of cattle-based control measures alone."*

On the basis of these findings, and subsequent independent advice in 2007, the Welsh Government plumped *for* a targeted cull of Badgers in one specific area of west Wales, whereas the UK Government vetoed that option in England and instead set up Badger vaccination trials in six areas of South-west England. By 2011, following elections in the two nations that brought to power two new governments, the positions were reversed. In England a four-year Badger cull was introduced in 2013 in parts of Gloucestershire and Somerset and five vaccination trials were cancelled. In 2015, the pilot cull was extended to parts of Dorset. In Wales, a four-year vaccination programme was launched in 2012.

Elsewhere in the UK, Scotland has long been free of bTB, so Badger culling is not an issue. In Northern Ireland, bTB peaked in 2002 but has subsequently decreased. The reasons are not known, but cannot be due to culling Badgers, as that has never happened there. The Northern Ireland Assembly started a hybrid culling and vaccination trial in 2014 called

Left: Badgers are a significant potential source of bTB infection in cattle – but there is no consensus on how best to address this.

Test and Vaccinate or Remove. Under this trial Badgers are caught and tested for bTB; those that are negative are vaccinated and released, those that are positive are removed/culled. Across the border in the Republic of Ireland, levels of bTB have remained broadly flat since the 1960s. The Irish government aims to maintain Badger populations through culling at less than one-fifth of their original densities over about one-third of the country's agricultural land.

Vilify, vacillate or vaccinate?

In Britain any decision to kill wild animals as a matter of policy is expected to be controversial. Culling Badgers has proved no different. This is an emotive issue, provoking strong reactions from all sides. Those in favour believe that culling is justly part of a wider strategy to tackle bTB. Many opponents believe that culling such a popular protected mammal is an expensive distraction from more effective and publicly acceptable measures. A petition against Badger culling on the UK Government website attracted the largest level of support for any such petition in the 2010–2015 Parliament.

Above: Opponents of Badger culls argue that vaccinating Badgers is a more effective, sustainable and publicly acceptable solution.

Many are worried about how effective culling actually is. There is good evidence that culling destabilises Badger social groups so that individual Badgers disperse. This increase in Badger movements appears to spread bTB infection in Badgers, and to raise the chance of infected Badgers coming into contact with cattle. This is known as the 'perturbation effect'.

Whichever side of the fence one sits, the common ground appears to be that vaccinating Badgers should have a role to play in preventing further spread of the disease. Charities such as the Badger Trust point to field trials of an injectable vaccine that reduced by three-quarters the proportion of Badgers testing positive for bTB. Badger vaccination is being used in a core area for bTB in Wales. In England the Government launched a scheme in 2014 to encourage vaccination in some areas around the edge of the main bTB area. In both countries vaccination was paused in late 2015 due to a global shortage of the vaccine. All parties – industry and charities alike – are lobbying for speedy production of a vaccine for cattle. The shared intention is that the UK's cattle, Badgers and other animals will all be bTB-free.

The RSPB's view

The topic of bTB is a complex and emotive issue that cannot be covered in its entirety within these pages. Here is a summary of the RSPB's perspective on Badger culling.

The RSPB is sympathetic to the concern in the farming community over the impact that cattle control measures can have both financially and emotionally on individual farmers. However, the RSPB does not believe that the Badger culling is the way forward and has not permitted Badger culling on its land.

The RSPB believes that culling Badgers risks making bTB worse unless it is carried out in a coordinated and highly synchronised way, over extensive areas, and for at least four years. There is considerable risk that such a cull would be impracticable and unsustainable. Independent assessment of shooting free-ranging Badgers suggests that this approach, as operated in the 2013 pilots in England, failed to meet effectiveness and humaneness targets.

Vaccinating Badgers does not risk making bTB worse. The RSPB believes that a combination of cattle testing and movement controls, biosecurity measures and vaccination is a more sustainable, publicly acceptable solution to the problem of bTB.

For further information on this issue a general summary is available on the UK Parliament website http://researchbriefings.parliament.uk/ResearchBriefing/Summary/SN05873#fullreport

In addition see the Further Reading section on page 125.

Badger Fun

Since the 20th century, Badgers have popped up frequently in European literature. One of the first writers to feature Badgers in books was the British author Kenneth Grahame. He was also the first to buck the prevailing trend of treating Badgers as foes.

The much-loved character of Mr Badger is one of the five core animal protagonists in Grahame's justly celebrated 1908 story, *The Wind in the Willows*.

Our first meeting with Mr Badger cleverly captures a typical experience with a wild Badger. Mr Badger peered out from the seclusion of a hedge, 'trotted forward a pace or two, then grunted "H'm! Company," turned his back and disappeared.' Consistent with nature, Badger lives deep in the woods and prefers not to mix with other animals. As Rat observes, 'Badger hates Society, and invitations, and dinner, and all that sort of thing'.

Opposite: In the film version of Roald Dahl's book *Fantastic Mr Fox*, the Badger is called Clive.

Below: Badger (top left) with Mole, Ratty and Mr Toad in ITV's adaptation of *The Wind in the Willows*, which ran to five series between 1984 and 1990.

Above: Aeron Clement, a Welsh science-fiction writer, wrote a *Watership Down*-style story about Badgers entitled *Cold Moons*.

When disturbed by Rat and Mole at his home, Badger is initially grumpy, but soon becomes contrastingly relaxed, hospitable and accommodating. Badger shuffles along the meandering tunnels of his sizeable underground home, slumbers gladly and is loathe to venture outside in winter. Grahame is clear about the importance of a subterranean lifestyle to Badgers. 'There's no security, or peace, except underground,' says Badger. 'Up and out of doors is good enough to roam about and get one's living in; but underground – that's my idea of *home*!'

Granted, Grahame treats himself to some artistic licence in his depiction of Badger – rather than feed Hedgehogs, real Badgers routinely feed *on* them – but his sympathies for Brock shine through the tale. The author celebrates Badger as a figure of natural authority, a beacon of calm and stability, a rock for other animals less assured or rooted. Grahame honours the resilience of Badgers and their fidelity to their homeland, despite the incursion of humans. 'People come – they stay for a while – and they go,' says Badger. 'But we [Badgers] remain. There were badgers here long before that same city ever came to be. And now there are badgers here again.'

Tommy Brock, and more besides

A few years after Grahame's tale was published, another British author, Beatrix Potter, provided a rather different take on Badgers. In *The Tale of Mr Tod*, Potter recounts the relationship between 'two disagreeable people', the eponymous Mr Tod (a Fox) and Tommy Brock (a Badger).

Potter's portrayal of Tommy Brock is far from flattering. He is a 'short bristly fat waddling person... not nice in his habits', perpetually dirty and a scrounger of baby Rabbits.

Subsequent fictional representations were closer to Grahame's sympathetic portrayal. In *Prince Caspian* by C. S. Lewis, Trufflehunter courageously helps Caspian regain the throne and thereby return Narnia to its former glory. The best friend of comic-strip character Rupert Bear – whose exploits first graced the *Daily Express* in 1920 – is Bill Badger.

The theme of Badger as dutiful sidekick – Sancho Panza to Don Quixote – also emerges in Roald Dahl's *Fantastic Mr Fox*. Dahl does not give his Badger a name, but in the 2009 film of the same name, director Wes Anderson raises his prominence by calling him Clive. Voiced by well-known comic actor Bill Murray, Clive is both Mr Fox's lawyer and his best friend.

The duo survive a feisty relationship, with Clive's loyalty frequently stress-tested by Mr Fox's antics and Mr Fox considering his solicitor buddy 'too respectable'. In one scene, Clive fails to dissuade his friend from buying a tree-home in an area renowned for the persecution of Foxes. Mr Fox should have heeded Clive's advice, for he is soon ambushed, his tail shot off, and his home destroyed. The sidekick as sage...

Above: Comic actor Bill Murray voiced Clive the Badger in the film *Fantastic Mr Fox*.

Below: In the film adaptation of *Fantastic Mr Fox*, Clive the Badger is Mr Fox's lawyer and best friend.

Henry Williamson's best-known work is *Tarka the Otter*, but an earlier tale in his *Animal Saga* was *The Epic of Brock the Badger*. The 'epic', tragically, was the baiting of an embattled old boar. In *The Animals of Farthing Wood* by Colin Dann, Badger is a respected, considerate peacemaker who leads the collective of creatures when Fox is absent. Occasionally, Badgers become protagonists, as in the *Grandville* series of graphic novels by Bryan Talbot, first published in 2009. The lead character is Detective Inspector LeBrock of Scotland Yard, a Badger armed with bullets.

In J. K. Rowling's fabulously successful *Harry Potter* series, the emblem of the house of Hufflepuff is a Badger. Rowling characterises Hufflepuffs as patient and dependable, and as stoic defenders of friends and family. Does this sound familiar…?

Badgers in teaching children

Above: The St John's Ambulance run a Badger Programme for children that teaches first-aid and other life skills.

Ascribing positive characteristics to Badgers has encouraged equally positive associations in the world of health, safety and worldly compassion. In the UK, for example, the St John's Ambulance runs a Badger Programme for children between five and ten years old. In addition to learning basic first aid techniques, youngsters learn a variety of 'subjects', such as caring for animals and protecting the environment. Also in the UK, the Royal Society for the Prevention of Accidents launched its 'Tufty Club' in 1953. Tufty – a Red Squirrel – and Sergeant Badger have taught road safety to generations of British children, this author included.

Right: Reproduced by kind permission of The Royal Society for the Prevention of Accidents; the junior branch, The Tufty Club, is fronted by Sergeant Badger.

Left: Badgers serve as mascots for sporting teams, including the Fulham football club in England, and the Wisconsin Badgers in the United States.

Badgers add colour to our lives by serving as mascots for sports teams. 'Billy the badger' is the mascot for English football club Fulham. Billy has occasionally courted controversy, including once being sent off by the referee for dancing on the pitch. In the United States, an American badger named 'Bucky' is the official mascot of the University of Wisconsin–Madison. He attends major sporting events for the Wisconsin Badgers athletics team.

Watching Badgers

In many places seeing a Badger is easy. Just drive along dual carriageways and you will surely zoom past the sad sight of a chunky, silvery, stripe-faced corpse. We may despair of ever seeing a Badger with a heart thumping as resolutely as our own. Yet such defeatism is misplaced. Seeing a *live* Badger is actually surprisingly easy.

There are three main ways to intentionally watch Badgers (as opposed to luckily bumping into one while out at night), and all involve being close to a sett as twilight encroaches. The first and most challenging thing is to discover your own sett, piecing together clues in the countryside that betray the mammal's presence.

The second way is to visit a known sett after talking to local landowners, naturalist friends or local Badger groups. Viewing facilities are what you yourself make of them: a tree for shelter, a bush for disguise or a cushion for comfort. Thirdly, you can pay to join an organised outing to an established watchpoint over a sett. Sitting in the comparative comfort of a hide, floodlights may illuminate the 'stage', and frequently habituated Badgers may gobble peanuts within centimetres of your awestruck face. Each option has its merits; I have tried and enjoyed them all.

Opposite: There are various ways to watch Badgers, but pretty much all involve sitting quietly at a sett as dusk falls and Badgers think about emerging.

Below: Keep alert for evidence of Badgers, such as this hole, while out and about in the countryside.

Recognising Badger holes

The key clues to whether you have found a Badger hole or that of another mammal lie in the sizes and shapes of the holes you find, as these fit the body of the mammal that uses them. Rabbit holes are typically roughly circular, and Fox holes are tall and narrow, whereas those of a Badger are shaped like a half-moon, with the flat side downwards. Badgers, being bigger than the other two mammals, need much larger holes – typically 20cm (8in) high and perhaps 30cm (12in) wide. In contrast, a Rabbit hole is roughly the diameter of a tennis ball.

For supporting evidence, respectfully peer inside the first foot or so of the tunnel. If the walls are rubbed smooth and there is neither evidence of bones from small mammals nor the distinctive pungent scent of a Fox, you have almost certainly located a Badger's domain. For unequivocal confirmation look for a sizeable and steep-sided spoil heap adjacent to the sett, and perhaps scattered clumps of dry vegetation that once served as bedding.

Countryside clues

For animals whose underground lifestyle is premised on concealment, Badgers deliberately advertise and inadvertently leave abundant signs of their presence. If you teach yourself what to look for, it is pretty easy to work out if your local neighbourhood, nature reserve or Sunday stroll route hosts Badgers. So what should you be looking for?

The biggest and most obvious indication, of course, is the sett. A large, bare patch of earth, on a slope, with sizeable holes is likely to be the home of a group of Badgers. Likely, but not certain. The Red Fox and Rabbit can have superficially similar homes, so you will need to check a few details before you can be sure.

Once everything points to your discovery being a Badger sett, extend your search to some other characteristics. Can you find a scratching tree with the bottom metre of its bare trunk testament to regular and repeated clawing? Have a go at identifying the species of tree; the chances are that it's an Elder. Then you might scour the ground for the shallow pit of a latrine, where calls of nature also serve as signals ('message in a bottom'?). You might also check for snuffle holes, where soil or vegetation has been rooted up in the search for food. Or squint around for dead wood torn apart in the hunt for woodlice. Just take care not to spend too long around the sett; too strong a human scent will distract and potentially disturb the mustelid occupants.

Above: Conical pits containing faeces are evidence of a Badger latrine, which means you are in a Badger territory.

Below: Bare ground below wood-land and scratched trees are good indications that Badgers are present.

Above and below: Generations of Badgers have trodden traditional paths across the countryside, and they don't let obstacles like dry-stone walls deter them.

Badgerways

Another telltale landscape feature is the strikingly clear paths ploughed by Badgers. These are so regularly used, often over many years, that they furrow through pasture or other vegetation, even planted crops. So traditional are these 20cm (8in) wide Badger thoroughfares ('Badgerways'?) that animals may belligerently continue to use them even if a physical barrier is built in their way. Even a house may not deter a Badger; it will simply round the obstacle, relocate the well-worn, scent-marked path and continue. Follow a Badgerway and, if you guess the right direction, it will lead you to the sett.

Above: Badgers often lose hairs to barbed wires fences as they squirm underneath. Each coarse hair is distinctively black, white and grey.

There are two great ways to check whether the path you are investigating was created and is being used by Badgers. The first method is to follow the trail until it crosses a barbed-wire fence. As a Badger squirms underneath, the twisting barbs may trap one or more back hairs. Long, coarse and coloured black and white, these can come from no other mammal.

The second way is to check the ground for pawprints. Look for a muddy or snowy area, and check carefully. A Badger print is wider than it is long. The central pad is shaped rather like a kidney bean. Either four or five toes will make an impression, and each will usually have a

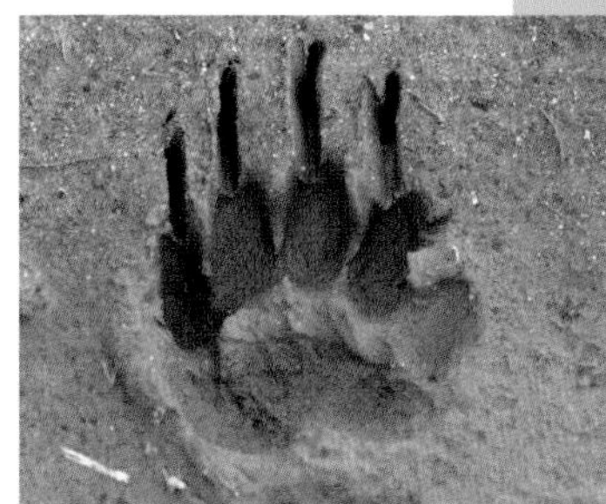

Above and left: Badger pawprints are distinctive; the pad being wider than it is long, with a clear imprint of four or more claws.

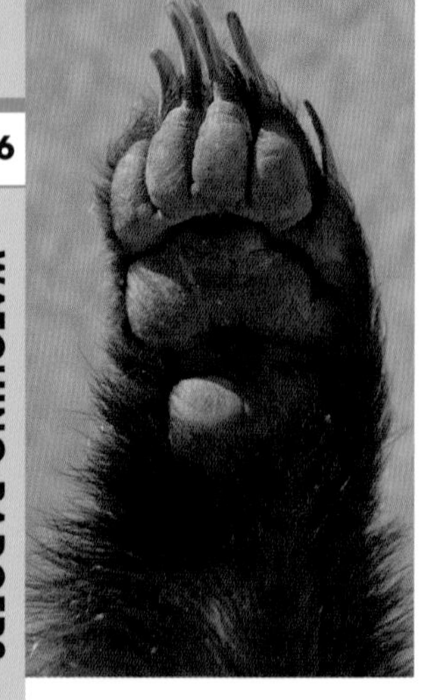

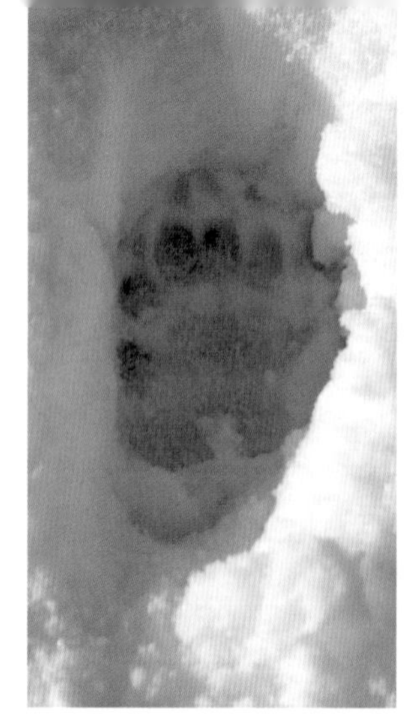

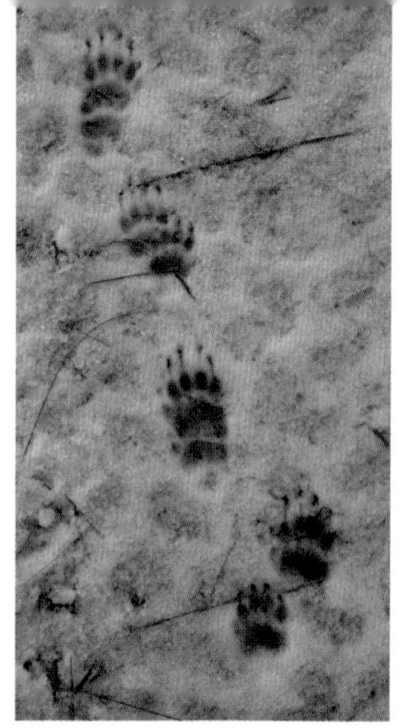

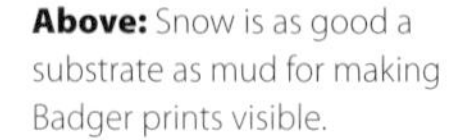

Above: Snow is as good a substrate as mud for making Badger prints visible.

narrow claw mark. Note, however, that Badgers often place their rear foot where their front one was, which smudges the tracks.

Poo!

Earlier in the book we learnt how important Badger toilets are for communicating messages, particularly near the sett and along the boundary between neighbouring territories. Latrines are typically shallow, cone-shaped depressions rootled into the soil, generally 5–20cm (2–8in) deep. The contents are left uncovered – it makes no sense to conceal the signals they are offering. The size and shape of the poo depends very much on the Badger's recent meals. A supper of Lobworms may result in a soft, squidgy spiral, whereas the fruit- and berry-rich diet of autumn may result in rather jelly-like deposits.

Below: Badgers dig cone-shaped pits in which to poo; examining the contents reveals what the animal has eaten recently (in this case, millet).

The Badger business

'Build it and visitors will come' is the motto of ecotourism entrepreneurs. In many places in the UK some people with Badgers on their land have grasped the potential commercial value of enabling the general public to enjoy watching these mammalian mint humbugs. There is also at least one offering in the Belgian Ardennes. The typical visitor proposition involves watching Badgers from one or more hides. These are often dug into the ground, so that visitors witness Badgers at eye level – making the experience delightfully intimate. The viewing area is typically baited, with peanuts or other food laid out to encourage the Badgers to stay longer, in full view and at close range. Viewing conditions are sometimes enhanced by artificial lighting, to which the Badger group has become accustomed.

Left: The Badger sett below the Margaret Grimwade hide in Suffolk, from which the adjacent two photographs were taken.

Above and below: At well-established Badger-watching hides to which the public is welcome, peanuts or other food is typically scattered on the ground to encourage the mustelid residents to stay around.

The Badger Trust, a British charity, has compiled a list of commercial and charitable Badger-viewing operations across England, Scotland and Wales. This can be downloaded from its website (see p. 125). Here are a few personal favourite locations.

At College Barn Farm in Oxfordshire, Richard and Sandra Butt rent out a secluded mobile home that comes with private, effectively 'en suite', Badger watching. The heated hides even have flaps for photography. Income generated is ploughed back into conservation on the farm. The set-up is similar at Old Henley Farm in Dorset. In Scotland, Aigas, Cairngorm Wildlife and Speyside Wildlife operate Badger-watching hides. Even guest houses and hotels have cottoned on to the attraction, with residents gawping at Badgers on the patios of accommodation from the Lake District to Cornwall, and from Pembrokeshire to the Isle of Wight.

Since 2005, the Badger Trust has acted as an umbrella organisation for around 80 local Badger groups across Britain. Many of these organise Badger-viewing excursions for their members. The Lancashire Badger Group, for example, has its own purpose-built hide overlooking a Badger sett in the Ribble Valley. The Trust's website provides the groups' contact details: give your local organisation a call to learn about viewing opportunities.

During the summer the Suffolk Wildlife Trust – in collaboration with the Suffolk Mammal Group – offers guided viewings of Badgers three times a week at its

Margaret Grimwade Badger Hide, set on private farmland near Ipswich. Monies raised (and it is an admirable moneyspinner, being booked out throughout the summer) help the Trust with its conservation programmes. The Leicestershire and Rutland Wildlife Trust runs similar events at its Rutland Water reserve, as does the Scottish Wildlife Trust at Falls of Clyde reserve. At Wakehurst Place in Sussex, the Royal Botanic Gardens, Kew, has a hide dedicated to Badger watching in its Loder Valley reserve.

Choosing how to watch Badgers

There are clear advantages in viewing Badgers using guides and hides. You get instant action, at close range. Appearances are near guaranteed (but not entirely, because blank nights are not unknown) and usually prolonged. The Badgers are used to human voyeurs, even in groups, and are sometimes downright exhibitionist. Floodlit conditions facilitate photography, and you can watch in comfort, sheltered from the elements and perched on a seat comfier than the ground.

There are downsides, too. You will be largely free of the exhilarating tension of not knowing whether or not you will see the Badgers. You will miss the cautious emergence of naturally wary animals. With a floodlit 'stage' and pane

Below: Keeping still, quiet and downwind of the sett are key to any successful Badger-watching experience.

Above: Badgers are nocturnal creatures; this image was taken using an infrared remote camera trap.

of glass separating watcher and watched, even views at an ostensibly intimate distance may leave you feeling disconnected. You may not experience the privilege of having the Badgers to yourself – and you will not truly join, if only temporarily, the Badger in its environment.

Above: Using binoculars at dusk helps you 'see in the dark'.

Below: Couple a torch with a red filter (or red plastic) to avoid disturbing Badgers.

Seeing in the dark

Not being creatures of the night, our eyes are not particularly suited for night vision. So how on Earth can we actually watch Badgers given that they are out and about after dark? Fortunately, human vision is better adapted than we might think. It is amazing how quickly your sight can adjust to even the darkest night. Squinting out of the sides of your eyes may help, as it optimises use of the specialist light-gathering properties of the rod cells.

Using good-quality binoculars can be a boon, as they capture light efficiently and can extend viewing by an hour or so. If you plan regular visits to setts, you might consider investing in affordable night-vision equipment, typically called night scopes, or night monoculars. What about torches? Surprisingly, some Badgers seem not to be distracted by the sudden use of a flashlight, or indeed fixed floodlights at a commercial venue. To be safe, however, fixing a red filter or simply a piece of red plastic over the light would enable you to watch without alerting the watched.

Badger-watching tips

Understanding how Badgers use their senses is key for wannabe watchers to observe without being detected. Knowing Badger habits is similarly important: there's next to no point in turning up at a sett in mid-afternoon or halfway through the night.

- Badgers have a hypersensitive sense of smell and can detect human interlopers from a considerable distance. Accordingly, check the wind direction before you go out. **Always approach and sit downwind of a sett.** Avoid wearing scent or smelling of the curry house where you went for dinner. Importantly, make sure you have a wee before you go out!
- Badgers are alert to sudden movement. An uncomfortable, fidgeting watcher is likely to disturb them. **Choose a sitting position that you can sustain without needing to stretch.** A cushion may be useful.
- Badgers start at unfamiliar noises. **Rustle-free garments are a must: wool and fleece are good**, puffer jackets and cagoules are not. Avoid coughing, sneezing, whispering or reaching into your pocket for a sweet.
- Badgers may have poor eyesight, but they readily discern pale colours. **Wear dark clothing.** Exposed skin glows in the moonlight. If possible, cover up your hands with gloves and your face with a scarf or balaclava.
- Badgers can also spot shapes that shouldn't be there. **Break up your silhouette by making use of your surroundings.** You could sit by or behind a tree or bush, or even climb a tree.
- The chances of Badgers smelling, hearing or seeing you reduce the further from the sett you sit. Clearly, there is a balance between remaining undetected and being so far from the action that you can't really see anything. A rule of thumb is to **stay at least 10–20m (30–60ft) from a sett.**
- Badgers like mild, wet nights, as these encourage Lobworms to surface. Don't be deterred by a bit of rain.
- **Badgers are not fans of cold breezes.** If the wind is easterly or northerly, harbour low expectations.
- Time your arrival carefully, taking into account the season. **For most of the year Badgers typically emerge from the sett at or after dusk.** In summer, however, it can be much earlier – and still completely light. You will want to be in place an hour or so before you expect the first activity.
- If you visit a sett by day, perhaps to do a recce, **your scent may make the inhabitants reluctant to emerge** that night. Best to return another evening for a viewing.
- Finally, **ensure you have permission to be at the sett.** Where relevant, check with the landowner that they are happy with your visit.

An encounter with Badgers

It is early evening in late April. Flanked by Friesian cows browsing placidly, I weave my way along a footpath dissecting green pastures. A Skylark performs operatics overhead. Dunnocks flirt and fossick in the seclusion of a hedgerow. As I reach the broadleaved woodland and enter it, the stillness of pre-dusk seeps between trunks and envelops me. A Blackbird breaks the silence with a scolding rattle. Whether or not the call is directed at me, it reminds me to stay unobtrusive in terms of sound, sight and smell.

Wood Anemones star the woodland floor, and the lilac haze of Bluebells is complemented by the mineral coolness of their perfume. A few score metres ahead I spy bare earth on a slope. Pausing to double-check that I am downwind of this miniature terracotta bluff, I spot a Gorgonzola of holes which confirms that I have reached my destination. Secreting myself at the base of a suitable tree, I make myself as comfortable as the hard ground and roots permit. I become still.

As the gloaming first encroaches, then pervades, I attune myself to these sylvan surroundings. A Pheasant squawks distantly, a cow lows and Woodpigeons explode from the canopy without apparent reason. The mild air heightens the wooded scents, intoxicating even to my impoverished sense of smell. I glimpse a Roe Deer padding its way along a ridge. It is unaware of me. I take that as a good sign.

I train my eyes on the Badger sett. Which hole to look at? With every minute it is getting darker, demanding that my eyes work ever harder. With every minute I trust my eyes less. With every minute tension mounts. Will they or won't they? Will I or won't I?

Then – a flicker of white, the merest hint of a movement. Perhaps I am imagining things, my desire to see Badgers overwhelming the evidence. Then my ears detect strange noises; a whickering answered by a terse grunt. Another brief wave of white handkerchief against the black of a burrow. Surely this is for real?

Above: It is dusk: an adult Badger sniffs above ground then emerges, tentatively, from its sett.

Then the black lump of coal of a Badger nose bobs up, indisputably and incontrovertibly. A zebra-striped snout emerges, sniffs the air and retreats. A minute later it repeats the process – but this time, the Badger emerges. It is an adult male, or boar, a tank of a mammal. Grizzled and hunchbacked, the boar snuffles around, squints towards me and has a good scratch. Then he signals to the whole Badger world that he has survived another day, sinking his bottom to the ground and anointing it with scent.

Another Badger emerges, slightly but discernibly smaller; a female, or sow, I suspect. She mooches over to the boar and grooms him briefly. The male rewards his hairdresser by scent-marking her, confirming her ongoing membership of the group. She plods off along a leaf-free path, a gleaming 'Badgerway' through the profound obscurity of the forest. Her tail surprises me with its length and fluffiness. I think, bizarrely, of a stunted Old English Sheepdog.

Above: Be in place before dusk at a Badger sett and familiarise yourself with the lie of the land.

Then the highlight; what I've been hoping for. Another adult Badger emerges, swiftly followed by three bonsai versions. Cubs! Mustelid Mini-me, in triplicate. The youngsters appear nervous: I may be a privileged witness to an early excursion into their brave new world. After 15 minutes the darkness is comprehensive, and I realise that I have lost sight of the Badgers. Whether they have shuffled off into the night or retreated underground, I do not know. But they are gone, and this realisation hastens my own departure – smiling and sated.

Above: Once she is sure that the coast is clear, the female Badger (or sow) brings her cubs above ground.

Glossary

boar Male Badger.

carnivore Meat eater.

cub Juvenile Badger.

dispersal Movement of animals away from their birthplace to become established elsewhere.

genus Taxonomic level above species and below family; it forms the first word in a species' scientific name.

meline Like a Badger.

metabolism Speed at which an animal converts food into energy.

mustelid Member of the weasel family (Mustelidae).

nocturnal Active mainly at night.

omnivore Animal that eats both meat and plant material.

pelage Hairy or furry coat of a mammal.

predator Species that feeds on a specified animal.

reintroduction Where captive-bred or captive-raised animals are returned to the wild.

sett Badger's den.

sow Female Badger.

territory Area actively defended by an animal against others of the same species.

torpor Prolonged period of inactivity.

vermin Animal considered a pest and thus often legally killed.

Acknowledgements

Mike Unwin kindly introduced me to Bloomsbury. There, Julie Bailey, Jane Lawes and Jasmine Parker were wonderful editors. Sharon Lowen commented with insight on an early version of the text. Patrick Barkham kindly allowed me to quote sections from his inspirational book *Badgerlands* and explained where to see Badgers near my Norfolk home. Ed Barnham-Jones welcomed me to a sett at Pensthorpe Natural Park, Norfolk. I also enjoyed watching Badgers at the Margaret Grimwade Hide near Ipswich, courtesy of Michael Strand at Suffolk Wildlife Trust and Adrian 'Marmite' Hinchcliffe of the Suffolk Mammal Group. John Dixon, Adrian Hinchcliffe and Dougal Macneil inspired me with their Badger experiences. Jan Hein van Steenis advised on Badger watching in Belgium. Caroline Mead, Robin Wynde and Kate Smith (RSPB) commented on sections of the text. Joe Dawson and Yoav Perlman allowed me to take photographs of them in the field. Finally, without the support of my wife Sharon and our little cub Maya, this book would never have been written.

Further Reading and Resources

Fiction

Dahl, Roald, *Fantastic Mr Fox* (George Allen & Unwin, 1970)

Dann, Colin, *The Animals of Farthing Wood* (John Goodchild, 1979)

Grahame, Kenneth, *The Wind in the Willows* (Methuen, 1908)

Lewis, C. S., *Prince Caspian* (Geoffrey Bles, 1951)

Potter, Beatrix, *The Tale of Mr Tod* (Frederick Warne & Co, 1912)

Talbot, Bryan, *Grandville* (Dark Horse, 2009)

Williamson, Henry, *The Epic of Brock the Badger* (Macdonald, 1960)

Non-fiction

Badger vaccination – Carter S. P., Chambers M. A., Rushton S. P., Shirley M. D. F., Schuchert P., *et al.* (2012) BCG Vaccination Reduces Risk of Tuberculosis Infection in Vaccinated Badgers and Unvaccinated Badger Cubs. *PLoS ONE* 7(12): e49833. doi:10.1371/journal.pone.0049833

Barkham, Patrick, *Badgerlands* (Granta, 2013)

Bourne J., *et al.* 2007 *Bovine TB: the scientific evidence*. http://webarchive.nationalarchives.gov.uk/20090330154646/www.defra.gov.uk/animalh/tb/isg/pdf/final_report.pdf

Childs, Keith, *The Badger Diaries* (Bookworm Publications, 2010)

Clark, Michael, *Badgers* (Whittet Books, 1988)

Heath Justice, Daniel, *Badger* (Reaktion Books, 2015)

Independent Expert Panel (2014) Pilot badger culls in Somerset and Gloucestershire https://www.gov.uk/government/uploads/system/uploads/attachment_data/file/300382/independent-expert-panel-report.pdf

Jenkins H. E., Woodroffe R., Donnelly C. A., (2010) The Duration of the Effects of Repeated Widespread Badger Culling on Cattle Tuberculosis Following the Cessation of Culling. *PLoS ONE* 5(2): e9090. doi:10.1371/journal.pone.0009090

Neal, Ernest, *The Natural History of Badgers* (Croom Helm, 1986)

Neal, Ernest & Chris Cheeseman, *Badgers* (T. & A. D. Poyser, 1998)

Pearce, George, *Badger Behaviour, Conservation & Rehabilitation* (Pelagic Publishing, 2011)

Roper, Timothy, *Badger* (HarperCollins, 2010)

Spencer, Adam, 'One Body of Evidence, Three Different Policies: Bovine Tuberculosis Policy in Britain, *British Journal of Politics and International Relations* 31(2), 2011

Woods, Michael, *The Badger*, 2nd edn (The Mammal Society, 2010)

Yalden, Derek & Stephen Harris, eds, *Mammals of the British Isles: Handbook*, 4th edn (The Mammal Society, 2008)

Resources

The RSPB www.rspb.org.uk

The Wildlife Trusts www.wildlifetrusts.org

The Mammal Society www.mammal.org.uk

Defra http://tinyurl.com/DefraBovineTB

Natural England www.gov.uk/badgers-protection-surveys-and-licences

Partnership for Action Against Wildlife Crime http://tinyurl.com/GovPAW

The Badger Trust www.badgertrust.org.uk

Badgerlands www.badgerland.co.uk

The Badger Friendly Directory www.badgerfriendlyfarms.info

Scottish Badgers www.Scottishbadgers.org.uk

Badger Protection League www.badgerprotectionleague.com

Save the Badger www.savethebadger.com

National Farmers Union www.nfuonline.com/science-environment/bovine-tb

Badger watching in Belgium www.landschapvzw.be/fotografiehutten

Image credits

Bloomsbury Publishing would like to thank the following for providing photographs and permission to reproduce copyright material.

While every effort has been made to trace and acknowledge all copyright holders, we would like to apologise for any errors or omissons and invite readers to inform us so that corrections can be made in any future editions of the book.

Key t = top; l = left; r = right; tl = top left; tcl = top centre left; tc = top centre; tcr = top centre right; tr = top right; cl = centre left; c = centre; cr = centre right; b = bottom; bl = bottom left; bcl = bottom centre left; bc = bottom centre; bcr = bottom centre right; br = bottom right

AL = Alamy; FL = FLPA; G = Getty Images; NPL = Nature Picture Library; RS = RSPB Images; SH = Shutterstock

Front cover t Eduard Kyslynskyy/SH, b Stanislav Duben/SH; **spine** Stanislav Duben/SH; **Back cover** t Klaus Echle/NPL, b Andrew Mason/FL; **1** Andrew Mason/FL; **3** Dietmar Nill/Minden Pictures/FL; **4** Klaus Echle/NPL; **5** t The Wildlife Trusts, b Adrian Hinchcliffe; **6** DEA/M. SEEMULLER/G; **7** l Erik Mandre/SH, r Paul Hobson/FL; **8** tl Ian Rentoul/SH, tr Hugh Clark/FL, b Mike Lane/FLPA; **9** t James Lowen, c Flip de Nooyer/Minden Pictures/FL, b James Lowen; **10** t James Lowen, cl Biosphoto, Guy Piton/Biosphoto/FL, cr Gerard de Hoog, NiS/Minden Pictures/FL, bl Gerard Lacz/FL, br Mark Sisson/FL; **11** Andy Rouse/NPL; **12** tr Roland Seitre/NPL, tl Nick Garbutt/NPL, bl Donald M. Jones/Minden Pictures/FL, br Jurgen & Christine Sohns/FL; **13** t Konstantin Mikhailov/NPL, bl Clinton Moffat/SH, br Prue Simmons; **14** t Andrew Mason/FL, b Bertie Gregory/2020VISION/NPL; **15** t Paul Hobson/NPL, b Felix Man/Stringer/G; **16** tl Richard Costin/FL, tr Danny Green/RS, bl Sebastian Kennerknecht/Minden Pictures/FL, br Geography Photos/G; **17** t Adrian Hinchcliffe, bl Richard Costin/FL, br Adrian Hinchcliffe; **18** Andy Rouse/NPL; **19** Elliott Neep/FL; **20** tl Oliver Smart, tr James Lowen, b Adrian Hinchcliffe; **21** t Biosphoto, Fabian Bruggmann/Biosphoto/FL, b Adrian Hinchcliffe; **22** tl Paul Hobson/FL, tr Adrian Hinchcliffe, c Hugh Clark/FL, b Christian Cabron/Biosphoto/FL; **23** t Adrian Hinchcliffe, c Konstantin Mikhailov/NPL, b UniversalImagesGroup/G; **24** t Chien Lee/Minden Pictures/FL, b Adrian Hinchcliffe; **25** t Klaus Echle/NPL, c Dietmar Nill/Minden Pictures/FL, b Elliott Neep/FL; **26** l Adrain Hinchcliffe, r Dietmar Nill/Minden Pictures/FL; **27** t Flip de Nooyer/Minden Pictures/FL, c Alex Hyde/NPL, b Adrian Hinchcliffe; **28** Andrew Mason/FL; **29** Fred Hazelhoff/Minden Pictures/FL; **30** Adrian Davies/NPL; **31** Barry Turner/AL; **32** Adrian Hinchcliffe; **33** t BSIP/G, b Andrew Mason/FL; **34** Simon Colmer/NPL; **35** l Barcroft/G, r Martin B. Withers/FL; **36** t Andrew Cooper/NPL, b Adrian Hinchcliffe; **37** t Adrian Hinchcliffe, b Kevin J Keatley/NPL; **38** t Martin B. Withers/FL, b Adrian Davies/NPL; **39** t Andrew Mason/FL, c Andrew Cooper/NPL, b Bertie Gregory/2020VISION/NPL; **40** Robert Canis/FL; **41** Kevin J Keatley/NPL; **42** t Fabrice Cahez/NPL, b Paul Hobson/NPL; **43** t Andrew Mason/FL, b S Charlie Brown/FL; **44** t John Eveson/FL, b Derek Middleton/FL; **45** t Michael Durham/FL, c Oliver Smart, b Robert Canis/FL; **46** t Mike Lane/FL, b Klaus Echle/NPL; **47** Adrian Hinchcliffe; **48** t John Hawkins/FL, b Adrian Hinchcliffe; **49** t Photofusion/G, c Stanislav Duben/SH, b Colin Varndell/NPL; **50** James Lowen; **51** tl Adrian Hinchcliffe, tr Oliver Smart, bl Adrian Hinchcliffe, br Adrian Davies/NPL; **52** t Andrew Parkinson/2020VISION/NPL, b Adrian Hinchcliffe; **53** t Adrian Hinchcliffe, b Adrian Hinchcliffe; **54** Derek Middleton/FL; **55** James Lowen; **56** t James Lowen, bl JMN/G, br Olaf Protze/G; **57** t Laurent Geslin/NPL, b Rod Williams/NPL; **58** tl Klaus Echle/NPL, tr Paul Hobson/FL, b Adrian Hinchcliffe; **59** Adrian Hinchcliffe; **60** tl James Lowen, tr Eye Ubiquitous/G, cl DEA/CHRISTIAN RICCI/G, cr Adrian Hinchcliffe, b Adrian Hinchcliffe; **61** l Paul Hobson/FL, tr James Lowen, br IMAGEBROKER, RICHARD DORN/Imagebroker/FL; **62** t DEA/G. NEGRI/G, b Meul/ARCO/NPL; **63** t IMAGEBROKER, ALESSANDRA SARTI/Imagebroker/FL, b James Lowen; **64** t S Charlie Brown/FL, b Adrian Hinchcliffe; **65** t Klaus Echle/NPL, bl Kevin J Keatley/NPL, br Klaus Echle/NPL; **66** t Paul Hobson/NPL, bl Michael Krabs/Imagebroker/FL, br Roger Hosking/FL; **67** t Nigel Cattlin/FL, c Adrian Hinchcliffe, b Andy Rouse/NPL; **68** t Sean Hunter/FL, bl Adrian Hinchcliffe, br Kevin J Keatley/NPL; **69** l James Lowen, r James Lowen; **70** tl Adrian Hinchcliffe, tr Adrian Hinchcliffe, b Andrew Mason/FL; **71** t Michael Krabs/Imagebroker/FL, b Sue Robinson/SH; **72** John Eveson/FL; **73** Jasper Doest/Minden Pictures/FL; **74** tl Oliver Smart, tr John Gooday/Nature in Stock/FL, b Eduard Kyslynskyy/SH; **75** t Birdphoto/Nature in Stock/FL, tc James Lowen, bc James Lowen, b Imagebroker/FL; **76** t James Lowen, bl Oliver Smart, br Oliver Smart; **77** Tom Marshall/NPL; **78** t Paul Hobson/FL, bl Photofusion/G, br James Lowen; **79** Photofusion/G; **80** t VYACHESLAV OSELEDKO/Stringer/G, bl Adrian Hinchcliffe, br Colin Seddon/NPL; **81** t Colin Seddon/NPL, b Adrian Hinchcliffe; **82** Klaus Echle/NPL; **83** t CARL COURT/Stringer/G, b Barcroft/G; **84** t Oli Scarff/G, b Sebastian Kennerknecht/Minden Pictures/FL; **85** l Tony Hamblin/FL, r Adrian Hinchcliffe; **86** MyLoupe/G; **87** Terry Whittaker/NPL; **88** t feiyuezhangjie/SH, tc Kiev.Victor/SH, bc Oleg Golovnev/SH, b Robert Canis/FL; **89** Adrian Hinchcliffe, b Dan Kitwood/Staff/G; **90** t Mar Photographics/AL, b AFP/Staff/G; **91** l Jeremy Early/FL, r Sam Hobson/NPL; **92** t David Hosking/FL, b David Hosking/FL; **93** l Mike J Thomas/FL, r Adrian Hinchcliffe; **94** t Michael Durham/FL, b Adrian Hinchcliffe; **95** Owen Brown/SH; **97** Andrew Parkinson/NPL; **98** Oliver Smart; **101** Mark Bowler; **102** tl Terry Whittaker/NPL, tr Terry Whittaker/NPL, bl Tom Marshall/NPL, br Tom Marshall/NPL; **103** Andrew Parkinson/2020VISION/NPL **104** Vera Anderson/G; **105** Moviestore Collection Ltd/AL; **106** Terry Smith/G; **107** t Ian Gavan/Stringer/G, b Stephen Lovekin/Staff/G; **108** t Ian Patrick/AL, b ROSPA; **109** David Cannon/Staff/G; **110** Elliott Neep/FL; **111** James Lowen; **112** tl Claire Plumridge/SH, tr Dominique Delfino/Biosphoto/FL, cl Dar1930/SH, cr Roger Wilmshurst/FL, bl James Lowen, br Sean Hunter/FL; **113** t James Lowen, bl James Lowen, br Oliver Smart; **114** tl Adrian Hinchcliffe, tr James Lowen, bl Tim Graham/G, br James Lowen; **115** tl Andrew Parkinson/NPL, tr James Lowen, bl Oliver Smart, br Cisca Castelijns, NiS/Minden Pictures/FL; **116** tl Jean E. Roche/NPL, tc Adrian Hinchcliffe, tr Michael Durham/FL, bl Adrian Hinchcliffe, br Adrian Hinchcliffe; **117** l Adrian Hinchcliffe, tr James Lowen, br James Lowen; **118** t James Lowen, b James Lowen; **119** Colin Seddon/NPL; **120** t Eric Medard/NPL, c James Lowen, b James Lowen; **121** James Lowen; **122** David Kjaer/NPL; **123** t James Lowen, b Andrew Mason/FL; **124** Bertie Gregory/2020VISION/NPL

Index